No ordinary gardener

Thomas Knowlton, 1691–1781

No ordinary gardener

Thomas Knowlton, 1691–1781

Blanche Henrey

EDITED BY A. O. CHATER

British Museum (Natural History)

First published 1986
BM(NH)Cromwell Road
London SW7 5BD

British Library Cataloguing in Publication Data

Henrey, Blanche
No ordinary gardener Thomas Knowlton, 1691–1781.
1. Knowlton, Thomas 2. Gardeners—Great Britain—Biography
I. Title II. Chater, A. O. III. British Museum (Natural History)
635′.092′4 SB63.K5/

ISBN 0-565-00976-1

Printed in England by
St Edmundsbury Press Ltd, Bury St Edmunds, Suffolk

Contents

Preface

For more than 30 years Miss Blanche Henrey worked almost daily in the Botany Library of the British Museum (Natural History), with a degree of meticulous care and sustained effort that culminated in the publication in 1975 of her definitive *British botanical and horticultural literature before 1800.* Her scholarship and distinctive contribution to the life of the Museum was recognised by the Trustees, who appointed her an Honorary Associate in 1973. In the years preceeding her death in 1983, she was working on a biography of Thomas Knowlton based mainly on original sources in the Museum Library. This work goes beyond a mere description of Knowlton's life, interesting though that will be to gardeners, and provides a vivid illustration of the lively debate and thirst for new information that he shared with the men of science of the time. Despite valiant efforts in the face of increasing physical weakness, Miss Henrey was not able to bring the manuscript into a state ready for publication before her death. Fortunately, Mr A. O. Chater, a senior member of the Department's staff, with whom she had discussed the project on many occasions, was willing to undertake the final revision of the typescript for the press. This proved to be a considerable task, involving much checking of numerous extended quotations from Knowlton's letters with the original manuscript sources, which are in a difficult hand with many confusing abbreviations. We are grateful for the care which he has brought to this exacting task.

In discussing her manuscript with the Museum Publications Officer, Miss Henrey had offered some funds in support of publication, and we are grateful to her surviving sister and to her executors for facilitating this intention after her death. We are also grateful to the Stanley Smith Horticultural Trust, through its Director, Sir George Taylor, himself an old friend of Miss Henrey of many years' standing, for the grant of £1000 towards publication costs. Mrs B. A. Bakerswell also wished to contribute towards the publication in memory of a long-standing friendship.

Blanche Henrey was an amiable and unassuming colleague, whose natural modesty cloaked a remarkable knowledge of British horticulture and natural history, at the time when the foundations of modern science were being laid. The Museum is fortunate to have had her as a devoted Associate, and we take special pleasure in the publication of this, her final work.

J. F. M. Cannon
Keeper of Botany

Acknowledgements

The author left no formal list of acknowledgements, although many of those who provided information or helped to elucidate particular problems are acknowledged in the relevant places in the text. She would certainly have wished to acknowledge the much wider help given by many of her colleagues, contacts and correspondents, only a token few of whom are known to the editor and mentioned here. At the British Museum (Natural History) her colleagues in the Department of Library Services, including Mr M. J. Rowlands, Mr A. P. Harvey, the late Miss P. I. Edwards, Mrs J. A. Diment, Mrs C. North, Mrs G. A. Ballance and Miss L. Newington provided continual encouragement and support. In the Department of Botany Miss C. A. Whitefoord helped especially in interpreting pre-Linnaean names, and Mr R. Ross, Dr E. Launert and Mr P. W. James were among the many others who helped on numerous occasions. The support of Dr M. Nickson and her colleagues at the British Library was greatly valued, and was particularly helpful in the early stages of the project. Other archivists and librarians throughout the country answered questions and provided documentation without which much of this book could not have been written, among them Mr P. J. Day and Mr M. A. Pearman of the Devonshire Collections, Chatsworth, Mr K. D. Holt of the Humberside County Archives, Miss A. G. Warrilow of the Bodleian Library, Oxford, and Miss C. Boddington of the University of Hull Library. Permission to reproduce manuscripts from the relevant libraries is gratefully acknowledged. Dr D. J. Mabberley read an early draft of the manuscript and helped in many other ways, and among others whose encouragement and assistance was repeatedly given were Mr D. Neave, Dr and Mrs I. Hall, Dr C. H. Brock, Mr D. McClintock, and Mr J. H. Harvey. The author would especially have wished to acknowledge the encouragement and support given by her sister, Miss K. H. Henrey. Mrs T. Smith typed most of the manuscript, Mrs A. Surrey compiled the index, and Miss M. J. Short gave valuable help with the proof-reading. Errors and inconsistencies in the text must be the responsibility of the editor, as these would certainly have been eliminated by the author had she been able to undertake the final revision of the manuscript.

List of letters in Part two

Letters addressed by Knowlton to:

Dr Patrick Blair	12 January 1727/28
Samuel Brewer	12 April 1728
	23 October 1728
Dr Richard Richardson	22 April 1729
Samuel Brewer	3 July 1729
	24 August 1729
Sir Hans Sloane	20 September 1729
Dr Richard Richardson	29 October 1729
Samuel Brewer	17 November 1729
	8 February 1729/30
	28 March 1730
	12 May 1730
	[n.d.]
	16 June 1730
	8 July 1730
	22 August 1730
	28 June 1733
	27 August 1736
	1 October 1736
	22 October 1736
Dr Richard Richardson	31 October 1736
	6 November 1736
Samuel Brewer	5 July 1737
	August [1737]
	9 October 1737
	26 October 1737
	7 February 1737/38
	6 July 1738

	13 August [1738]
	13 September [1738]
	28 September [1738]
	3 October 1738
	18 or 19 November [1738]
'Dr Richard junior'	
(Richard Richardson, Esq.)	18 November 1739
Peter Collinson	20 January 1740/41
Letters addressed to Knowlton by Henry Simpson, the Earl of Burlington's personal secretary:	11 July 1741
	25 July 1741
	8 August 1741
	8 August 1741
	30 January 1741/42
	3 July 1742
	17 July 1742
	26 August 1742
	13 October 1742
	23 October 1742
	28 October 1742
	9 July 1743
Letters addressed by Knowlton to:	
Samuel Brewer	15 December [1741]
Sir Hans Sloane	27 December 1741
Richard Richardson, Esq.	8 June 1743
	18 July 1749
	29 September 1750
	8 March 1750/1
John Ferret The Earl of Burlington's London agent	22 March 1750/1
Letter addressed to Knowlton:	
by Henry Simpson	27 March 1742

Letters addressed by Knowlton to:

Richard Richardson, Esq.	5 July 1751
	1 October 1751
	2 June 1752
	24 April 1753
William Stonehouse, Gardener	27 January 1754
Richard Richardson, Esq.	21 July 1754
William Stonehouse	5 November 1754
Peter Collinson	2 February 1755
William Stonehouse	23 February 1758
Richard Richardson, Esq.	9 March 1758
	13 November 1758
William Stonehouse	27 February 1763
Richard Richardson, Esq.	May day [1763?]
John Grimston	28 May 1763
Emanuel Mendes da Costa	19 July 1765
Peter Collinson	13 July 1767
John Ellis	October 1770
John Grimston	11 January 1777
	27 March 1777
	24 July 1779

Introduction

Knowlton in the context of the eighteenth century

It is unusual to be able to find details concerning so much of the life of an eighteenth century private professional gardener as in the case of Thomas Knowlton. Born in 1691, he was the eldest of the five sons of William and Ann Knowlton of Chislehurst, in Kent. Of his father nothing is known. Neither can any information be found about Thomas Knowlton until 1711, when he was 20 years of age. From that date, however, until shortly before his death in 1781, material is available from letters and other manuscripts, as well as from several publications, concerning the life, work, and interests, of this intelligent, gifted, and interesting personality.

Knowlton was undoubtedly one of the most outstanding gardeners of his time. His advice was constantly sought as a horticulturist, garden designer, and planter, especially in Yorkshire, and his skill in the cultivation of rare and exotic plants was widely recognized. Furthermore, he was noted for his interest in our native flora and was responsible for several first records of plants in the British Isles. Richard Pulteney, in his *Historical and biographical sketches of the progress of botany in England*, **2**, 239 (1790), wrote of Knowlton a few years after his death:

> 'His zeal for English botany was uncommonly great, and recommended him successfully to the learned botanists of this country'.

Six years later, in 1796, the botanist Richard Anthony Salisbury named a genus of Ranunculaceae in honour of Thomas Knowlton, basing *Knowltonia* on a South African species which Linnaeus in 1753 had named *Adonis capensis* and which is now known as *Knowltonia capensis* (L.) Huth. (The genus now contains eight species and is confined to South Africa[1].)

Knowlton's interests and enthusiasms were not limited to his work as a botanist and gardener. His general curiosity and desire for knowledge extended to several other fields of study, including antiquarian pursuits. In

(1) *See* Rasmussen (H.), The Genus *Knowltonia* (Ranunculaceae), *Opera Botanica,* **53** (December 1979).

this he reached a position of some note, as is evident from his correspondence, and from the number of his letters read before various meetings of the Fellows of the Royal Society. His correspondence of antiquarian interest includes the finding of a medieval jug near York, the discovery of the foundations of a Roman Road that went through Londesborough Park in East Riding, the Roman remains unearthed near the same property at Millington and considered at the time to be the site of the Roman station of Delgovicia, and the finding at Cowthrop in Yorkshire, of the skull and horns of the Great Deer, or Irish Elk, *Megaloceros giganteus*, considered probably the first of this species to be found in England.

Other fields of natural history, besides botany, in which Knowlton showed an interest, were ornithology, and the study of fossils. He assisted at various times the botanist Samuel Brewer in his search for star stones – the stem ossicles of fossil crinoids – found at Bugthorpe and Leppington, and also for trochites – the wheat-like joints of the stem of fossil crinoids – found in the district of Craven in West Riding. Furthermore, the eminent author and fossilologist, Emanuel Mendes da Costa, sought Knowlton's assistance in getting him some belemnites, then said to be found in plenty in parts of East and West Riding.

Although it became the fashion in the eighteenth century for all those with any pretensions to culture to possess a library, it is unlikely that most professional gardeners were owners of more than a very small collection of books, consisting perhaps of a few popular works on practical gardening by such well known authors as Stephen Switzer, Richard Bradley, and Philip Meles. Occasionally, gardeners were fortunate enough to be permitted to use books from the libraries of their masters. Knowlton must have been unusual for he collected a large amount of literature on botany, gardening and horticulture, as well as on a variety of other subjects. Unlike most gardeners he seems to have had no financial problems.

During the many years that Knowlton worked at Londesborough there were long periods when his master, the Earl of Burlington, was not in residence, and Knowlton was able to visit other Yorkshire landlords who had sought his advice. For example, he gave assistance to Sir Charles Hotham with a new design for a garden and paddock at his property at South Dalton, near Beverley. He helped Thomas Duncombe when a beautiful landscape was being designed at Duncombe Park, overlooking Rievaulx Abbey. On one occasion Knowlton wrote to Samuel Brewer telling him that he had promised to go to the Honourable Thomas Willoughby of Birdsall Hall, near North Malton. All these estates were in

Yorkshire. Beyond that county his advice was sought by Zachariah Bourryau regarding 'some new water works' for Blyborough Park, Lincolnshire.

Knowlton gives very little information concerning the part he actually played in assisting owners in the design of their gardens and parks. It is likely that among the manuscript material preserved in private houses and in public institutions some very interesting details might come to light. For instance, in the Humberside County Record Office, at Beverley, are preserved receipts for goods purchased, plans made, and work carried out, for William Constable at his estate of Burton Constable in the Holderness area of East Riding, Mrs Ivan Hall of Beverley, who has been studying this manuscript material, found a number of documents connected with Knowlton. Thus, with her help it has been possible to add to the little information given by Knowlton himself concerning what he did at this estate.

There is one Yorkshire property, however, of which there is much information concerning the part Knowlton played in the planning and planting of the garden and park. This is Everingham, then Sir Marmaduke Constable's estate, near Londesborough. Knowlton spent a great deal of time over a period of more than ten years visiting Everingham, in order to give advice and practical help, as well as providing, when required, young tree seedlings. This information has been found in the Constable of Everingham estate correspondence now preserved in the Library of the University of Hull. The present writer has transcribed passages from these letters whenever Knowlton's name is mentioned, and as these quotations provide a valuable record of the work carried out by Knowlton at Everingham during the years 1730 to 1743, they have been placed together in chronological order, according to date, in an Appendix (see p. 275).

The work that Knowlton thus undertook undoubtedly earned him a considerable amount of money, extra to the wages he received as gardener to the Earl of Burlington; some of this would have enabled him to purchase the books he wanted. He not only built up a fine library, but he also invested his money in real estate, and it is interesting to note that in legal documents he became described as 'Thomas Knowlton, gent.'

A subscriber himself to many publications issued in parts, Knowlton was tireless in assisting authors to get new subscribers. Two of those he helped in this way were the naturalists Mark Catesby and George Edwards. Among books he received as gifts were the large folio volumes of Sloane's *Voyage to ... Madera, Barbados ... and Jamaica* (1707–25).

The eighteenth century saw the introduction into Britain of many new and rare plants from foreign lands, and Knowlton's letters provide us with a lively picture of the enthusiasm and impatience with which he awaited publications with plates and cultural details of plants and trees that had not been previously illustrated. Although he seems usually to have known when a new publication was due to appear, he invariably asked a friend, such as Samuel Brewer, who lived at Bierley, near Bradford, and was in touch with many other naturalists, whether it was well thought of, and whether they themselves had seen it. Moreover, he seldom failed to criticize the text and plates, sometimes both, if he did not consider a work good. Even though Knowlton lived in a little village in East Riding, at the foot of the Wolds, there appears to have been little difficulty in the delivery of books, even such weighty tomes as the two folio volumes of Dillenius's *Hortus elthamensis* or the two Sloane folio volumes mentioned above. Neither did Knowlton appear to have problems in obtaining foreign publications.

Among the various early works used by Knowlton in the identification of plants were English and foreign herbals, Leonard Plukenet's *Phytographia*, *Almagestum botanicum*, published during 1691–1705, John Ray's *Historia plantarum* (1686–1704), and the same author's *Synopsis* of 1724, edited by J. J. Dillenius. It seems likely that Knowlton began his botanical activities at a very early age. In 1711, when he was 20, he started to write down what he considered were some of the more uncommon plants in Hertfordshire. Over a period of many years, until 1727, he continued to make notes of such plants in Hertfordshire, Bedfordshire, the island of Guernsey, and Yorkshire. In January 1728, he sent to Dr Patrick Blair, at that physicians request, a list of 31 plants he had thus noted. When the letter was sent Knowlton had only just moved to Yorkshire, where he remained for the rest of his life. In August 1729 he gave Samuel Brewer a similar, but much enlarged, list of such plants, the additional species having been mostly found in localities in Yorkshire.

After becoming gardener to the Earl of Burlington in Londesborough, Knowlton made a living collection there of native plants and on one occasion in a letter of 1770 he wrote that 'chamaepericlymenum or Dwarf Honeysuckle [i.e. *Cornus suecica*]' did well with him as did 'many of ye raire English Plants'. Samuel Brewer also cultivated British plants and he and Knowlton frequently wrote to each other for plants they required. For example in 1730 after Brewer had asked for *Cypripedium calceolus*, Lady's Slipper, Knowlton replied that he had only one plant and that had failed to flower. It probably died out, as in February 1758 Knowlton wrote to Richard Richardson's gardener at Bierley, to know if he could help him with such a request. Whether Knowlton received a plant is not known, but it appears that this rare species may still have been growing in Richardson's garden in November 1758, when Knowlton wrote asking, 'pray how dose y^{e}Ldy Sliper florish with you?' In July 1767, Knowlton's son Charles sent his father *Rubus saxatilis*, Stone Bramble, a gift that must have given him great pleasure, as it was a species he had never seen before. The collecting and growing of native plants in gardens was very popular among some botanists in Knowlton's day. Knowlton, as we have seen, went in for this pastime at Londesborough, and so did Brewer, Richard Richardson, whose garden, according to Pulteney, was 'well stored with ... a curious collection of English plants', and James Sherard of Eltham in Kent.

Although nothing is known of Thomas Knowlton as a gardener until 1720 when, at the age of 29, he was in the service of Sir Henry Penrice of Offly Place, Hertfordshire, he must by then have been in a position of some authority. It is likely that he had begun his chosen career many years earlier, possibly at about the age of 12, as an apprentice to a Master gardener. His first important job was, without doubt, as a gardener to James Sherard at Eltham, Kent. Here, this London physician and apothecary created, during the second and third decades of the eighteenth century, one of the most celebrated gardens of the day, richly stocked with plants from many parts of the world consisting in the main of species that were rare or only lately introduced into this country. In 1732 this work gave rise to the publication of the *Hortus elthamensis.*

Knowlton left Sherard's employ sometime before 20 February 1725, after having been in his service from the end of 1720 or 1721. The years Knowlton was gardener at Eltham had given him valuable experience in the cultivation and management of exotic flowers and fruits. At Eltham were built a greenhouse and stoves, and in February 1721 a hotbed of tanner's bark was made. This garden must have been one of the first in this country to use tan for hotbeds for the cultivation of Pineapple plants, and for the preservation of tender exotics.

The person who is generally considered to have been the first to succeed in growing Pineapple plants in England to produce fruit is Henry Telende, gardener to Sir Matthew Decker of Richmond, Surrey, who in the cultivation and ripening of the fruit used a hotbed of tanner's bark. An account of Telende's care and management of Pineapples was published in the July 1721 issue of Richard Bradley's *General treatise of husbandry and gardening*, and it seems possible that Sherard read this account and, as a result, was inspired to make his own tan hotbed. Previous to 1719 or 1720, which is considered to be the time Telende's plants were first producing fruit, very few hotbeds in England were made of tanner's bark. By October 1724, William Sherard wrote that his brother's garden was in excellent state with 'ripe Coffee, ripe Papaia, Ananas, and a great number of plants not to be seen elsewhere in England', and indeed rarely to be seen today!

When William Aiton, Gardener at the Royal Botanic Garden, Kew, was preparing the *Hortus kewensis*, which appeared in 1789, Thomas Knowlton gave him 'a variety of useful information'. This Aiton used with Knowlton's name 'always annexed'. It consisted of a list of plants with the names of persons by whom they were first introduced into England, or to whom their cultivation was first attributed. These were as follows:

Vol. 1. 272–3 '*Celastrus lucidus* [now *Maytenus lucida* (L.) Loes.]...
Nat. of the Cape of Good Hope
Introd. 1722 by James Sherard, M.D. Knowlton inscr.....'

Vol 2. 42 '*Rheum ribes* [Currant-fruited Rhubarb]...
Nat. of the Levant.
Cult. 1724, by James Sherard, M.D. Knowlton's mss....'

Vol. 2. 72 '*Arbutus andrachne* [Andrachne]...
Nat. of the Levant.
Cult. 1724, by James Sherard, M.D. Knowlton's mss....'

Vol. 2. 163 '*Prunus lusitanica* [Portugal Laurel]...
Nat. of Portugal and Madeira.
Cult. 1722, by Mr. Thomas Fairchild. Knowlton's mss....'

One of Aiton's entries (Vol. **3**, 411) reads:

'*Euclea racemosa*...
Nat. of the Cape of Good Hope.
Introd. 1722, by Mr. Thomas Knowlton...'

The type species of *Euclea* was known long before J. A. Murray described the genus in the thirteenth edition of the *Systema vegetabilium*, 747 (1774). (Murray's single species *Euclea racemosa* was described solely by reference to the Dutchman, J. Burmann's description and illustration of 'Padus foliis Subrotundis, fructu racemoso', in his *Rariorum africanarum plantarum*... *Decas nono*, 238, pl. 84, fig. 1 (1739).) It is likely that Knowlton received material of this plant from the Cape in 1722 via Holland.

Knowlton wrote an 'Account' of experiments he made relating to the culture and propagation of Coffee trees. This was dated 'Petworth in Sussex, Feb. 4. 1725–6'. Subsequently James Douglas, the eminent London physician, considered it so useful that he included it in his work on Coffee of 1727. This 'Account', together with Knowlton's record of his visit to the Island of Guernsey in the summer of 1726 to study the history

William Aiton (1731–1793), worked at the Chelsea Physic Garden under Philip Miller and was gardener at Kew 1759–1793, where he gained a European reputation for the Royal garden and compiled Hortus Kewensis *which included most of the plants cultivated in Britain at the time.*

and cultivation of the Guernsey Lily there, are apparently his only surviving writings before he went to Londesborough at the end of 1726. From January 1728 until July 1779, only just over two years before he died, he wrote a very great number of letters, many of which are still in existence.

Francis Drake, Yorkshire antiquary and author, considered Knowlton, 'a sensible, intelligent and most creditable person'. Through his personality, correspondence, work, and interests, Knowlton came in contact with many eminent men of his day, and the transcriptions from his letters given in the present work provide us with much of historical and social interest.

Knowlton numbered among his friends the President of the Royal Society, Sir Hans Sloane, and many of the Fellows, including the following botanists and naturalists, some of whom were authors of note: Dr Richard Richardson, Mark Catesby, Peter Collinson, George Edwards, John Ellis, Emanuel Mendes da Costa, and Dr Patrick Blair. Knowlton himself, besides having the distinction of a number of his letters being read before the Fellows was, on one occasion, on 11 May 1749, introduced at a meeting of the Society by the Secretary, Dr Cromwell Mortimer.

It might appear strange that Knowlton was never invited – as far as we know – to become a Fellow of the Royal Society. But the admission fee and annual subscriptions would have been a large item even to a man in his financial situation, and he would seldom have been able to come to London for meetings.

It is likely that Knowlton first felt the excitement and interest in growing exotic flowers and fruits when he worked in the Eltham garden, and again experienced it when he was gardener at Cannons, the Duke of Chandos's estate in Middlesex, where Coffee, and such fruits as Pineapples, Avocado Pears, Papaws, Custard Apples, and Guavas, were grown. Later again, Knowlton tells us himself that he had never been so happy as when he was at Londesborough thanks to the Earl of Burlington 'the best of L^{ds} & fraints'. At first, however, there were no buildings in this garden where tender plants could be grown, and at a time when new species were pouring into this country from all over the world this must, indeed, have been a disappointment. The garden and park at Londesborough were entirely redesigned in Knowlton's time, and the making of plantations of trees and shrubs, both native and exotic, was the order of the day. The publication of John Evelyn's *Sylva* in 1664, and its various later editions, resulted in many thousands of trees being planted in Britain during the eighteenth century. Burlington and other landowners in Yorkshire were no exception in wishing to follow this fashion.

Until a stove was erected at Londesborough in October 1729, the plants Knowlton referred to in his letters were chiefly those that he had seen growing wild during various journeys to different parts of Yorkshire, such as when visiting a friend who lived at Fountains Abbey, or going to the Earl of Carlyle's estate near Malton, or when travelling to Sir Charles Hotham's property at South Dalton, or riding from Lythe along the sea shore to Whitby, or when going to Walling Fen, some miles west of Hull.

Although Knowlton continued to collect native plants after he had his stove, and to exchange them with other keen botanists, he increasingly referred in letters to species native of foreign countries and indeed by June 1733 Knowlton was able to offer Brewer thirteen different species including plants native of the West Indies, the western Mediterranean, South Africa, and China. But it seems that it was the ambition of every gardener to grow Coffee, Pineapples, Guavas, Papaws, and Plantains and Knowlton had chosen such a selection for his new stove in 1729. However, at first, he was unsuccessful and his plants died. Although a number of greenhouses, stoves, and hotbeds were constructed at Londesborough, he continued having difficulty with these species until in August 1736 he was able to write to Brewer that he had 'great plenty of young coffee trees and likewise papaws' which he propagated by cuttings. He offered his friends plants of both. By November 1736 he was able to offer Dr Richardson plants of Guavas and Plantains. But it was Pineapples, that most popular of fruits in the eighteenth century, that Knowlton grew so successfully. The Earl, who greatly commended his gardener's Pineapples, had them sent to him at Chiswick and London, when he was not in residence at Londesborough, as well as having gifts of them delivered to various friends and relations in Yorkshire.

Whilst Knowlton did not hesitate to ask for plants that he wanted himself – in a letter of October 1736 he included a long list of plants that he was hoping friends could supply him with – he was always very generous in giving from his own collections. For instance, after Dr Richardson built a new stove he sent Knowlton a request for any plants he could spare. Knowlton offered him many different species native of the West Indies, Mexico, India, South Africa, and various other places. Requests were sometimes received to replace plants that had died. A variety of reasons are found in Knowlton's letters explaining why plants were lost, including lack of knowledge concerning their correct cultivation, too much or too

Pineapple (Ananas comosus (L.) *Merr.*), *an extremely popular fruit in the eighteenth century. Lord Burlington greatly commended the Pineapples that Knowlton grew and frequently presented them to his friends and relatives.*

Papaya (Carica papaya L.), *which Knowlton first grew in James Sherard's garden at Eltham and in the Duke of Chandos's garden at Cannons, and later cultivated with much success in the stoves at Londesborough.*

little heat, and the same regarding watering. Drought offered problems and so did frosty weather. Boxes sent by carriers were often slow in reaching their destination, or carelessly treated and the plants arrived dead. Knowlton always stressed the importance of not sending plants in pots which could, and often did, break and cause damage. Finally, the specimens sent were occasionally so poor that there was little chance of their survival under any conditions.

One person from whom Knowlton received gifts of seeds was Peter Collinson, a member of the Society of Friends, who introduced so many North American plants into our gardens. Concerning the exchange of seeds and plants that went on between botanists and horticulturists at this time, Collinson, in a letter dated 15 December 1735, wrote the following passage to John Curtis, of Williamsburg, a correspondent who sent him many seeds from America(1):

> 'It is . . . very necessary to share amongst us the seeds that come annually from abroad it not only preserves a Friendly Society but secures our collections, for if one doues not raise a seed perhaps another does & if one looses a plant another can supply him by this means our gardens are wonderfully improved in variety to what they was twenty years agon.'

The author and collector Mark Catesby also supplied Knowlton with seeds and plants, sending him on one occasion in 1738 *Magnolia grandiflora*. Knowlton greatly assisted this naturalist by getting subscribers to his *Natural history of Carolina*. It is interesting to note that he suggested that if Dr Richardson would only buy this work, the author 'would some times ablidge him wth some forain seeds as they came to hand.' To a keen botanist this was a very tempting proposition and one that may well have sold a number of copies of the work.

Knowlton also received various seeds and plants from persons left unnamed in his letters. For instance, he was sent seeds on at least two occasions from Italy, and he had a cargo of seeds from Africa and Barbary. That they were sent from a specified country, however, did not mean that they were natives of that place. For example, Bloodberry, *Rivina humilis*, native of the West Indies, was sent from Italy.

In 1740, the Reverend William Stukeley, after a visit to the garden at Londesborough, wrote that the greenhouse contained 'innumerable, rare, and exotic plants'. Although to our knowledge only one plant, namely *Euclea racemosa*, was first introduced into our gardens by Knowlton,

(1) *See* Swem (E. G.), Brothers of the spade, correspondence of Peter Collinson . . . and . . . John Custis . . . 1734–1746, *Proc. Amer. Antiq. Soc., April 1948,* 17–201 (1949), p. 47.

several species were apparently being cultivated by him at an earlier date than is recorded in Aiton's *Hortus kewensis* and other works. For example, it appears that he had *Tigridia pavonia*, Tiger Flower, in flower in 1739 even though Aiton states that it was first introduced into this country about 1796, and J. E. Smith traces back *Catharanthus roseus*, Madagascar Periwinkle, as being cultivated by Knowlton before 1756, while William Curtis records that it was first grown here, by Philip Miller, in 1757.

The horticultural writer, Richard Bradley, in his advice on the education of a gardener, published in his *General treatise of husbandry and gardening,* **1**, 169 (1724), gave as one of the directions to those who wished to follow the art of husbandry and gardening, that they should 'take every convenient opportunity of conversing with ingenious men, as well in the art of husbandry as gardening, and to view their different ways of practice...' Knowlton always followed this precept and was never happier than when visiting gardens or when hearing about them, whether they were private, institutional, or commercial. Among some 40 owners of gardens mentioned in Knowlton's correspondence the following were well known for their collections, and for introducing new species of plants: Sir Henry Goodricke of Ribston, Yorkshire, William Blathwayt of Derham, Gloucestershire, John Blackburn of Orford, Lancashire, Lord Petre of Thorndon, Essex, Lord Wilmington of Compton Place, Sussex, Peter Collinson of Peckham, and later of Mill Hill, near London, the Earl of Northumberland of Syon House, Middlesex, the Earl of Bute who made a botanic garden at Luton Hoo, Bedfordshire, the Earl of Coventry of Croombe d'Abitot, Worcestershire, Richard Warner of Woodford, Essex, and Dr Richard Richardson of Bierley, Yorkshire.

Knowlton loved coming to London because, as he wrote to his friend Peter Collinson, it had so many entertaining pleasures to engage him, referring perhaps to the people he met, and the gardens he visited, to say nothing of the plants that he received as gifts. On one of these occasions in 1741, he went to the gardens of Lord Wilmington and Lord Petre where he received plants from their collections, he met Sir Hans Sloane, Peter Collinson, Mark Catesby, Philip Miller, Gardener at the Chelsea Physic garden, and Isaac Rand, the Demonstrator of plants at the same institution, he dined one day with James Sherard's widow, and he also visited Whitton Place, the Earl of Ilay's estate in Middlesex. (It is said by Loudon in his *Arboretum et fruticetum,* **1**, 58 (1838), that between 1720 and 1742 the Earl of Ilay 'Collected all the foreign trees and shrubs which, at that time, were to be procured either at home or abroad.') When Knowlton was in London in May 1749 he was 'very frequently' with Sir Hans Sloane, Dr

Richard Mead, Dr Thomas Stack, and Dr Cromwell Mortimer, Fellows of the Royal Society who were all eminent in the field of medicine. From them he received 'several seeds & other Favours'. He also met two other friends and Fellows of the Society, the naturalists Mark Catesby and George Edwards. From Dr Mead's favours Knowlton raised *Melia azedarach*, Chinaberry, 'which', he wrote, 'is a rare and scarce Plant.'

Knowlton visited and wrote about in his letters the Royal gardens at Kew and the Botanic garden at Cambridge. He was also knowledgeable concerning many of the eminent nursery gardens of his day. He was for many years a friend of the famous nurseryman Thomas Fairchild of Hoxton, who sent him to Holland in 1726. He was also a friend of the nurseryman and garden designer Stephen Switzer, and was in touch with the two Yorkshire firms of Telford of York, and Perfect of Pontefract in West Riding. Both these firms specialized in trees and shrubs. Two famous nurserymen near London were James Gordon of Mile End of whom Knowlton thought very highly, and Christopher Gray of Fulham. Knowlton's friends at his request always kept him well informed with up to date news of horticultural interest. Soon after he went to Londesborough he wrote to Dr Patrick Blair, 'I Live now a grate distance from any twon [town] & very seldom see any new's papers'. Under the circumstances it is remarkable how aware he was of everything that went on in the world of botany and horticulture.

In July 1767, both James Lee and William Aiton told Knowlton about the beauty of *Alstroemeria pelegrina*, Peruvian Lily, which was flowering that summer at the nursery garden called The Vineyard of the firm of Kennedy and Lee, at Hammersmith. Kennedy and Lee had introduced this species from Peru in 1753. Knowlton also knew well the two brothers, James Scot of Turnham Green, and Henry Scot of Weybridge; Henry had been the Earl of Burlington's gardener at Chiswick House. Both men specialized in Pineapples, the fruit Knowlton himself was so fond of growing.

Of foreign firms, Knowlton had a very high opinion of Willem van Hazen of Leyden. In 1751 he went over to Holland to buy plants from them and during his visit he found the gardens in that country 'abounding with such numerose varity of Plants' that he 'could not help purchasing many of y^{m}'. Unfortunately, due to delays on his return journey he suffered many losses, but he evidently brought a large number of plants back safely and sent Richard Richardson the younger several lists with names and prices of his 'Holland plants'.

Unfortunately, it is not known what sort of education Knowlton received although his great interest in books, love of letter writing, and knowledge of Latin, albeit limited, indicate that he went to a good school perhaps at, or near, Chislehurst, the Kentish village where he was born. He certainly realized the importance of a good education, as is evident from the fact that he sent his son Charles to the Free Grammar School at Coxwold, in North Riding, and subsequently to St John's College, Cambridge.

After the Earl of Burlington's death in 1753 Knowlton was employed successively by the third, fourth, and fifth Dukes of Devonshire. Unlike Joseph Paxton, the sixth Duke's famous gardener, who is remembered by posterity as the designer of the Crystal Palace, as a Director of the Midland Railway, as a member of Parliament for Coventy, and for being knighted by Queen Victoria, Knowlton did not make a lasting name. He does, however, deserve our attention and recognition for his letters and other manuscript material which, over a period of many years, give an insight into the work, travels, interests, and friends of a local gardener in the eighteenth century who possessed talents, zeal, industry, and a fine character.

Knowlton's life and letters

The quotations in the present work from letters by James and William Sherard have been taken from the transcriptions of these letters as published by J. Nichols in his *Illustrations of the literary history of the eighteenth century*, and not from the original letters themselves, which are preserved with the Richardson correspondence in the Bodleian Library, Oxford. All quotations from the letters written by Thomas Knowlton, however, have been copied by the compiler herself from the original letters.

Knowlton's spelling and style have been faithfully preserved, as to do otherwise would have been to lose much of their raciness, and at the same time would have given a false representation of their author. Knowlton's handwriting is often difficult to read, especially when he wrote in a hurry due to pressure of work or in haste to catch the post. He had no pretensions to literary style, and his spelling, much of which is quasi-phonetic, is unconventional. Capital letters or small letters are used indiscriminately for names and titles, and capitals are frequently used in sentences for the initial letters of nouns, adverbs, adjectives, or other parts of speech. This is especially the case for the initial letter of words beginning with the letter *l*. Little attempt is made at punctuation and one sentence follows another without a break. Abbreviations are especially frequent, a feature of which is the use of superior letters in contractions: y^{e}, y^{m}, y^{r}, y^{t}, w^{ch}, w^{t}, for the, them, your, that, which, what. A mark of contraction is sometimes used over a letter after which letters are omitted, as recoṁendation for recommendation. The apostrophe, a sign used to indicate the omission of one or more letters in a word, Knowlton invariably forgot to insert. He sometimes reversed the letters *er* after an initial letter, as *persume* for presume, and often wrote the words *one* and *on* when his intended meaning was on and one. These errors in spelling can be very confusing to modern readers of his letters. Passages omitted by the transcriber are indicated thus…

Some of the details of Knowlton's life have been obtained from published works, as well as from manuscripts. Furthermore, in order to add to the value and interest of the various persons, places, plants, and other subjects that are mentioned in the quotations transcribed from Knowlton's letters, the present writer has added at the end of these quotations further information on the subjects concerned, while any additions made in the actual transcriptions have been placed within square brackets. Thus the modern Latin names of the plants are given in square brackets after Knowlton's phrase names or descriptions; common names are always given if the species is native of Great Britain, and sometimes given if the species was introduced from abroad.

It may be said that until the time of Linnaeus no convenient method for naming plants, either for genera or species existed. However, with the publication in 1753 of Linnaeus's *Species plantarum*, when this celebrated botanist consistently used his binomial or two word system of botanical nomenclature for every known species, plant names became simple to use and to remember. Unfortunately, Knowlton never made use of this innovation. He referred to a plant by the name given to it by an early author, by its common name, or by a short description of his own. Plants referred to by Knowlton only by their generic names have generally been omitted by the transcriber; only when the specific name of the plant could be identified with a fair certainty has it been thought worth including in the present work. The English names of trees mentioned in the passages from the Constable correspondence in Appendix A, have Latin specific names added wherever possible, e.g. 'Sycamore [*Acer pseudoplatanus L.*]', but this cannot be done when more than one species of a tree is referred to by the same common name, such as the several species of Elm.

The four volumes of J. E. Smith's *English flora* (1824) are invaluable when tracing the names of British plants listed by Knowlton. In this, under the heading of the Linnaean binomial name of each species, followed by the common name, the synonyms of earlier authors are given together with references to their works. Any of these names may be first looked up in the index to the volumes. Subsequently, *The Flora of the British Isles* by Clapham, Tutin, and Warburg (1962) and the *Flora europaea* (1964–80) may be used to make an up to date check of the name of each species.

Tracing the modern Latin names of exotic plants mentioned by Knowlton was often a long and tedious business. This had to be done in a first class botanical library as many pre-Linnaean and later publications had to be used. These included eighteenth century garden catalogues, and floras of many different regions. It was often helpful to start with the following works listing synonyms: J. P. de Tournefort's *Institutiones rei herbariae* (1719), H. Boerhaave's *Index alter plantarum* (1720), and A. van Royen's *Florae leydensis prodromus* (1740). Among later works C. Linnaeus's *Species plantarum* (1753), Philip Miller's *Gardeners dictionary*, eighth edition (1768) in which the author used Linnaean binomials, H. E. Richter's *Codex botanicus linnaeanus* (1835), and the *Index kewensis* are all necessary tools.

At the end of the present work are several appendices, the first of which consists of transcriptions of passages from the Constable of Everingham estate correspondence, which give information concerning work carried out at this estate by Knowlton. The second appendix gives details regarding Knowlton as a book lover and collector, followed by a list of some of the publications to which he subscribed. A record of Knowlton's transactions in the property market is given in the fourth appendix, followed by an abstract of his will. Finally, immediately before the Index is a list of plants dealt with by Knowlton, and others, mentioned in this work, arranged in families.

In the eighteenth century the county of Yorkshire was divided into three administrative divisions, East, West, and North Riding, and these divisions have been referred to by name throughout the present work when giving locations of the various places mentioned. To have used the new county boundaries introduced 1 April 1974 implementing the Local Government Act, 1972, would have been confusing. By the new Act the old county of Yorkshire was done away with and, generally speaking, replaced with four new counties, Humberside, and South, West, and North Yorkshire.

PART ONE

September 1691 to December 1726

Gardener to Sir Henry Penrice, James Sherard and the Duke of Chandos. A visitor to the Island of Guernsey. Probably Gardener to the Duke of Somerset and later to the Dowager Duchess of Buckingham.

Thomas Knowlton, the eldest son of William Knowlton and Ann Stokes, was born at Chislehurst in Kent[(1)] where he was baptized 6 September 1691 at St Michael's parish church.[(2)] He died 28 November 1781 in his ninety-first year and was buried in Londesborough churchyard in the East Riding of Yorkshire.[(3)] Of his childhood nothing appears to be known. He probably early showed an inclination for the study of natural history and he certainly had a knowledge of our native flora by the time he was 20. This is evident from a letter of 12 January 1727/28 addressed to the botanist Patrick Blair. In this Knowlton cites localities in Hertfordshire where he observed in 1711 the following plants: 'Cotula foetidissima flore pleno', *Anthemis cotula* L., Stinking Mayweed, 'Dentaria major', *Lathraea squamaria* L., Toothwort, and 'Periclyminum foliis quercinis', *Lonicera periclymenum* L., Honeysuckle.[(4)]

It has not been possible to discover where Knowlton went to school or at what age he began his training as a gardener. The earliest reference concerning his chosen career is found in Eleazar Albin's *Natural history of English insects* (1720), in which plate 88 is dedicated 'To Tho: Knowlton gardiner at S[r] Henry Penrisces at Offly Place in Hartfordshire.' Sir Henry Penrice LL.D. (1677–1752) was Judge of the High Court of Admirality, Chancellor of the Diocese of Gloucester and Official of Middlesex. It is likely that Knowlton had been in his employ for some years before the publication of Albin's book as, by 1714, he was studying the wild flowers of Offley Park, which was situated near Hitchin. In that year he found that 'Nidus avis' *Neottia nidus-avis* (L.) Rich., Bird's-nest Orchid, grew plentifully in that part.[(5)]

Knowlton had four younger brothers, John, William, Samuel, and Richard,[(6)] and it is of interest to note that one of the subscribers to Philip Miller's *Gardeners dictionary* of 1731 was 'Mr John Knowlton, gardener to Sir Thomas Seabright'. In the above mentioned letter of 12 January 1727/28 to Patrick Blair, Knowlton recorded seeing, 'Asarum of Dod[oens] [*Asarum europaeum* L., Asarabacca]: In a close not far from S[r] Tho[s] Seabrights of Beachwood Hartfordshire 1716.' Furthermore, in a

(1) *See* below, p. 85, Letter to Patrick Blair, dated 12 January 1727/28, preserved in the Hunterian Library, in the University of Glasgow.

(2) Chislehurst Parish Registers (preserved in County Hall, Maidstone, Kent).

(3) It is incorrectly stated on the tombstone over Knowlton's grave that he was 'aged 91' when he died.

(4) *See* below, p. 85, Blair, *op. cit.* Entries 6.11.24.[bis]

(5) Blair, *op. cit.* Entry 29.

(6) Chislehurst Parish Registers.

wood belonging to Sir Thomas, he observed 'Lychmachea chamaenerion flore purpureo', *Epilobium angustifolium* L., Rosebay Willow-herb.[1] It seems likely that this John Knowlton was Thomas Knowlton's brother who was born 26 November 1693.

It is reasonable to assume that Knowlton began his gardening career after leaving school, possibly being first apprenticed to the head gardener with a good many years experience and holding a position of responsibility at Offley Park. On 27 June 1720 he got married at Aston, near Stevenage, in Hertfordshire, to Elizabeth Rice,[2] and not long afterwards he returned to his native Kent where he became gardener to James Sherard.

James Sherard (1666–1738) F.R.S., after a successful career as an apothecary in Mark Lane, London, purchased a country house at Eltham in Kent and there he created a garden which became one of the most famous in England for its collection of exotic and rare plants. In stocking this garden James was greatly assisted by his brother, the noted botanist Dr William Sherard (1659–1728) F.R.S.[3] At Oxford, William worked for a degree in civil law but he also found time for botanical pursuits. Subsequently, he travelled on the Continent for several years and during this time visited the Jardin du Roi in Paris and made the acquaintance of the Professor of Botany Joseph Pitton de Tournefort (1656–1718). He also advanced his favourite study by botanizing in Switzerland and Italy. In 1688 he was in Holland where he visited the Leyden garden and there met Paul Hermann (1645–95), the Professor of Botany who was to become a lifelong friend. William Sherard was to make two further journeys to the Continent before he left England in August 1703 to travel to Smyrna where he was to take up his duties as English Consul. He remained in Asia Minor for many years and did not return to England until Christmas 1717. During these years he continued to collect botanical specimens and to correspond with naturalists both in England and abroad. Sherard left Smyrna in November 1716 and planned to return home by way of Italy, France, and Holland. In Holland he visited Hermann Boerhaave (1688–1738), who in 1709 had been appointed Professor of the Botanical Garden at Leyden.

(1) Blair, *op. cit.* Entries 2, 16.

(2) Parish Registers for St Mary's Aston, deposited in the Record Office, County Hall, Hertford. *See* also Stocking (Rev. C. H. W.), *The history and genealogy of the Knowltons of England and America*, **10** (1897) New York.

(3) In giving details concerning William Sherard the writer has made use of the thesis entitled: 'Consul Sherard: amateur botanist and patron of learning, 1659–1728', by George Pasti Jr. This was prepared for Pasti's doctorate of philosophy in history, at the University of Illinois, August 10, 1950.

James Sherard (1666–1738), a successful London apothecary who later created one of the most famous of English gardens at Eltham in Kent, where Knowlton worked in the early 1720s.

A few months before William Sherard reached home James Sherard wrote a letter, dated 8 June 1717, to the naturalist Dr Richard Richardson of Bierley, Yorkshire, in which he told him(1) 'when my brother is in England, I shall hope to have a garden of our own.' Nineteen months later on 7 February 1718/19 William Sherard informed Richardson:(2)

'I received your last curious parcel of mosses, and ought to have thanked you for them before this; but, removing my books, &c. into chambers I have taken, till my brother's house he has bought at Eltham in Kent is fitted up, I put it off.'

The following year, by 30 January 1719/20, William Sherard was able to tell Richardson,(3) that he had been 'most of the time at Eltham.' As soon as the property was acquired James moved fast to improve the garden, and on 10 May 1720 William wrote to the doctor:(4)

'My brother... is busy in building his green-house and two stoves, one at each end; and has laid out another according to Dr Boerhaave's design...'

Although James Sherard kept up his business as an apothecary, much of his time was now devoted to his garden, and by 28 March 1721 he was busy at 'Eltham building another stove to answer that at the east end of his greenhouse.'(5) The number of plants at Eltham was rapidly increasing, regarding which William Sherard wrote to Richardson on 7 September 1721 as follows:(6)

"Tis a fortnight this day that my brother and I return'd from our excursion to Paris, by way of Holland... We brought many plants from Amsterdam and Leyden, besides what seeds were then ripe; and expect the rest as soon as fit to be gathered. What we sent from France are not yet arriv'd, but expect them every day... We brought also at Haerlem a great number

(1) Nichols (J.), *Illustrations of the literary history of the eighteenth century,* etc., **1**, 406 (1817).
(2) *ibid.* **1**, 355.
(3) *ibid.* **1**, 363.
(4) *ibid.* **1**, 366–67.
(5) *Extracts from the literary and scientific correspondence of Richard Richardson,* etc. [edited by D. Turner], 166 (1835) Yarmouth.
(6) *ibid.* 172, 174.

of bulbs, which will soon increase... My brother is at Eltham, busy in looking after his new acquisitions, and building new stoves.'

Sherard promised to let Richardson have some of the seeds when they arrived from France, as well as some of the plants the next summer when they had been increased. During the eighteenth century there was much exchanging of plants and seeds among keen gardeners and owners of gardens.

The exact date when Thomas Knowlton went to Eltham is not recorded nor is it known why he went to work for Sherard. It seems likely, however, that after he married he was looking for a new job in which to gain further experience in botany and horticulture. Moreover, he was undoubtedly an ambitious man keenly interested in his work. William Sherard was a well-known botanist and collector and James, also, was an enthusiastic botanist. James, according to a letter of 7 February 1718/19 written by William Sherard to Dr Richardson,(1) aimed 'at having all the English plants in his garden, or near his house.' Knowlton must have been delighted with his new appointment and in a letter dated 14 October 1721 we find William Sherard writing to Dr Richardson:(2)

'The plants you were so kind to send were taken care of by our own gardener, who is an excellent fellow, as fond of plants as his master.'

When William and James Sherard returned from their excursion to Paris, by way of Holland, they were accompanied by Johann Jacob Dillenius (1687–1747), a German botanist whom William Sherard had persuaded to leave Giessen in Upper Hesse (Oberhessen) and return with him to London.(3) Dillenius, a native of Darmstadt, was already noted as an author, having published in 1718 his *Catalogus plantarum circa Gissam sponte nascentium*, a work reissued in 1719 with an Appendix. Sherard was particularly struck with Dillenius's knowledge of lower plants. In 1724, Dillenius began making drawings and descriptions of new plants in the Eltham garden. He did this at the request of William Sherard.(4) At first the work progressed slowly, but as the number of species in the garden increased he continued with greater diligence, until finally the work was

(1) Nichols (J.), *Illustrations*, etc., **1**, 355 (1817).
(2) *ibid.* **1**, 376.
(3) *Extracts from... correspondence of Richard Richardson*, etc., 172 (1835).
(4) Dillenius (J. J.), *Hortus elthamensis*, etc., v(1732) (Preface).

Johann Jacob Dillenius (1687–1747), a German botanist brought to England by William Sherard. After writing his account of Sherard's garden, Hortus elthamensis, *he became the first Sherardian Professor of Botany at Oxford.*

completed. It was published in two large folio volumes in 1732 or 1733 under the title of *Hortus elthamensis seu plantarum rariorum quas in horto suo Elthami in Cantio coluit... Jacobus Sherard.* The 325 plates, with accurately drawn and engraved figures of plants, accompanied with descriptive text, provide posterity with a fitting tribute to the famous Kentish garden at Eltham and the men who made it. Knowlton's opinion concerning the work will be given later.[1]

In a letter of 24 July 1722, William Sherard told Richardson that his brother, in the company of Isaac Rand, the Demonstrator of plants at the Chelsea Physic Garden, of the Society of Apothecaries, had gone to search the coast of Sussex for plants. Afterwards, he talked of making 'a tour into Kent, to fetch plants into his garden.'[2] Some months later, on 23 February 1722/23, Sherard addressed another letter to Richardson,[3] when he gave him further details of the progress of the Eltham garden:

'My brother seldom comes to town; of late indeed he could not well, his gardener having been in Holland, and returned but last week, and now the season of sowing prevents him; he has built a very convenient house to the south of the large mulberry-tree, divided into two rooms, one for raising seeds on hotbeeds, the other for keeping plants in tanner's bark.'

Tanner's bark was usually the bark of the Oak which, after being ground into a coarse powder, was used in tanning leather. When the tanning principle had been wholly exhausted the bark, now called tan, was taken out of the pits.[4] It was then used by gardeners to make hotbeds as it produced a moderate and even heat which lasted for a considerable time. Miller in the first edition (1731) of his *Gardeners dictionary* wrote that hotbeds made with tan were brought into use in England during the early years of the eighteenth century for tender exotic plants and the propagation of Pineapples. Such hotbeds, however, had long been used in Holland and Flanders.

William and James Sherard continued to add to the collection of plants at Eltham. In August 1723, after William had returned from a visit to Holland he wrote and gave Richardson the following news:[5]

(1) *See* below, p. 133.
(2) Nichols (J.), *Illustrations,* etc., **1**, 381 (1817).
(3) *ibid.* **1**, 384.
(4) Definition of tan taken from Rees (A.), *The Cyclopaedia,* etc., **35** (1819).
(5) Nichols (J.), *Illustrations,* etc., **1**, 388 (1817).

'I have brought over for my brother's garden what plants could be spared out of the Amsterdam and Leyden gardens, besides some I bought for him elsewhere, and all the seeds that were ripe; but I returned a month too soon for seeds. You shall have part of them.'

Some weeks after William Sherard arrived back in England his brother went abroad where he remained until late the following spring. During his travels he visited France and Italy and wherever he went he collected seeds.(1) In a letter written 2 November 1723, during the time James was abroad, William told the doctor:(2)

'I heard lately from my brother from Marseilles... He sent me over a Catalogue of plants wanted in the Paris garden, taken out of his Catalogue he carried with him, containing above six hundred; so far Eltham outdoes the King's garden, which I hope will bring you up to see it,'

About a year later on 6 October 1724, William Sherard wrote again to Richardson about James's success in horticulture, saying:(3)

'His garden is in excellent state, ripe Coffee [*Coffea arabica* L.], ripe Papaia [*Carica papaya* L.], Ananas, and a great number of plants not to be seen elsewhere in England.'

But the success of the Eltham garden must have been largely due to the care bestowed on it by Thomas Knowlton who was soon to leave Sherard's employ. This we learn from a letter dated 20 February 1724/25 addressed by William Sherard to Richardson, who told the doctor:(4)

'My brother's gardener has left in a huff (which he will have reason to repent), and he has sent to Holland for another.'

James Sherard had difficulty in replacing such a skilled man and in writing from Eltham on 3 May 1725 he informed Richardson:(5)

(1) *ibid.* **1**, 391. W. Sherard's letter to Richardson, dated 18 January 1723/4.
(2) *ibid.* **1**, 390.
(3) *ibid.* **1**, 394.
(4) *ibid.* **1**, 396.
(5) *ibid.* **1**, 409.

'I thank you for your invitation and readiness to accompany me into the North; but my old gardener having left things in the utmost disorder, and my new one not understanding much of my garden, this pins me down, and obliges me not to stir from home this summer.'

We shall find that throughout this biography Knowlton always showed himself to be a conscientious and hard-working person, and it is difficult to believe that he left the garden at Eltham 'in the utmost disorder'. It seems more likely that Sherard wrote the above passage as a result of his annoyance at losing such an excellent gardener.

By 25 March 1725, Knowlton was working as Head gardener to James Brydges, Duke of Chandos (1674–1744). In 1713, Brydges came into possession of Cannons, near Edgeworth (Edgeware) in Middlesex.(1) Soon he began to make alterations and improvements and gradually the property became famous for the grandeur of its house and gardens. Thus Knowlton, from working in a botanist's garden renowned for its collection of rare plants, now found himself in the employ of an immensely wealthy man with large pleasure gardens. These consisted of terraces, parterres, grass plots, wildernesses, and ornamental waters. Furthermore, there were stew-ponds for fish, and further afield fruit and vegetable gardens, groves and walks. Descriptions of Cannons may be found in the works of contemporary authors. In 1722, John Macky wrote:(2)

'The gardens are very large and well disposed; but the greatest pleasure of all is, that the divisions of the whole being only made by ballustrades of iron, and not by walls; you see the whole at once, be you in what part of the garden or parterre you will.'

Macky also drew attention to the large kitchen garden. Another writer, Daniel Defoe in 1725, after praising the beauty of the mansion house, the design of the garden, and the size of the canals, recorded:(3)

(1) James Brydges acquired Cannons from his wife's uncle, Warwick Lake, who died in May 1713. *See* Baker (C. H. C.) and Baker (M. J.), *The life and circumstances of James Brydges, first Duke of Chandos,* etc., 114 (1949).

(2) [Macky (J.)], *A journey through England. In familiar letters from a gentleman . . . The second edition,* etc., **2**, 10 (1722).

(3) [Defoe (D.)], *A tour thro' the whole island of Great Britain . . . By a gentleman,* **2**, Letter 3, p. 11 (1725).

'The avenues and vistas to this house are extremely magnificent, the great walk or chief avenue is near a mile in length, planted with two double rows of trees, and the middle walk broad enough for a troop of horse to march in front; in the middle way there is a large basin or fountain of water, and the coaches drive round it on either side; there are three other avenues exceeding fine, but not so very large; the beauty of them all will double, with time, when the trees may be grown...'

Further details concerning the gardens at Cannons have been published by C. H. C. and M. I. Baker in their book entitled, *The life and circumstances of James Brydges,* etc. (1949), compiled from the Duke's letters and papers now preserved in the Huntingdon Library, California. From this book it is learned that the Duke received the help of various friends and agents in obtaining plants and seeds from abroad. From Mr Ashley of Barbados came Pineapples, *Ananas comosus* (L.) Merr., Cinnamon, possibly Wild Cinnamon, *Canella winteriana* (L.) Gaertner, a tree that grows generally in the West Indies, Coffee berries and Coffee plants, *Coffea arabica* L., and 'avigator pears', *Persea americana* Miller, Avocado Pear. Messrs Harriman of Leghorn supplied, among other things, Broccoli seed and Fennel, and Evergreen Oak acorns, *Quercus ilex* L.. From Jamaica arrived various fruits such as Papaws, *Carica papaya* L., Star Apples, *Chrysophyllum cainito* L., Custard Apples, *Annona reticulata* L., Guavas, *Psidium guajava* L., and Tamarinds, *Tamarindus indica* L., while from Lisbon were procured Watermelon seeds, *Citrullus lanatus* (Thunb.) Matsum. & Nakai. But the fruit the Duke liked best of all was the Pineapple, which began to be grown at Cannons in 1719,[(1)] and when he sent a special gift of this fruit to a friend he sometimes requested that the top he returned to Cannons to make a new plant. It seems likely that the knowledge and experience Knowlton gained in growing Pineapples at Eltham and at Cannons helped him towards achieving the great success with this fruit when he grew it many years later at the Earl of Burlington's garden at Londesborough in Yorkshire.

Little appears to be known of what Knowlton actually accomplished at Cannons although, in a letter dated 20 September 1729, he wrote to Sir Hans Sloane that when he worked for the Duke of Chandos he raised several seeds of the Logwood tree, *Haematoxylum campechianum* L., but 'leaving that place *persum* was all lost.' Sloane in his work entitled *A voyage to the islands Madera, Barbados, Nieves, S. Christophers and Jamaica*, etc., 183

(1) Henry Telende, gardener to Sir Matthew Decker of Richmond, was also growing Pineapple plants, and these were producing fruit about this time.

and pl. 231, fig. 1 (1725), wrote:

'Logwood is cut about the town of Campeche [Mexico] in great quantities, and brought to Jamaica in sloops to be sent into Europe, by the traders in Jamaica. It is ground to power by mills made for that purpose, and used by dyers, as I am told, to give their cloaths a good ground and foundation for other colours.'

Knowlton must have raised his seeds of these trees in a hothouse as the plants are tender and cannot be grown in England in the open ground.

It appears that Knowlton worked at Cannons for several months. Among the British historical manuscripts in the Huntingdon Library there is an account book in which half of one page is devoted to 'Thomas Knowlton. Head gardiner' who received six separate payments, totalling £86, for work carried out between 25 March and 30 September 1725. Five of the payments were for Knowlton's gardeners and labourers, while the sixth, a sum of £2.2s. 6d.,(1) was payment 'in full for tulips and all demands.' It is not known how many persons Knowlton employed but it probably varied from month to month. According to C. H. C. and M. I. Baker:(2)

'In a January week in 1725 three persons worked in the parterre and four in the north and greenhouse gardens. In a March week the same number were employed, with twelve more laying and raking gravel walks.'

Men were paid 1s. 3d. or 1s. 2d. a day; women 8d.

After Knowlton left Cannons he may have worked for a time at Petworth House in Sussex, when the owner was Charles Seymour, sixth Duke of Somerset. In a letter of 12 January 1727/28 to Patrick Blair, Knowlton recorded seeing several plants in the vicinity of Petworth, such as *Campanula persicifolia,* Narrow-leaved Bellflower, in 1725 'in plenty', when travelling from Petworth to Midhurst. Furthermore, in Douglas's work entitled *Arbor yemensis fructum cofe ferens: or, a description and history of the coffee tree,* 54–55 (1727) is an 'Account of the culture of the coffee plant' by Thomas Knowlton, dated at 'Petworth in Sussex, Feb. 4. 1725–6.' James Douglas (1675–1742), the eminent London physician and anat-

(1) In the old currency, abolished in 1971, the pound sterling was divided into 20 shillings (s.), and the shilling was divided into 12 pence (d.).

(2) *The life and circumstances of James Brydges, first Duke of Chandos,* etc., 161.

omist, and a keen and active botanist,[1] considered Knowlton's 'Account' would be valuable to those who wished to cultivate the plant in England, as it was written from the author's own practical experience in growing Coffee at Eltham. A transcription is given below:

'In the latter end of July, 1723... the worth Doctor Sherard (with whose brother I then liv'd) brought over from the Physick garden of Amsterdam, one Coffee tree of about three foot high, and one berry. The berry, carefully cover'd over with wax, was given to me, with some directions relating to the culture of it, from Cornelius Voss, gardiner at Amsterdam; but these being no ways agreeable to my own notion, nor to the practice of gardening here in England, I begg'd leave to use my own skill, without being confin'd to them. The berry being open'd, parted into two seeds, and having prepar'd a good fresh rich sandy soil, I put them into a small pot fill'd therewith, about two inches deep, and immediately after plung'd the pot into a temperate hot-bed. In about four or five weeks afterwards I had the pleasure of seeing one of them come up, with his cap on his head; and the other follow'd in a week afterwards. They continu'd growing very fast both winter and summer alike, and in twelvemonth's time were above two foot high.

Having now three trees in good health, I resolv'd to try some experiments relating to the culture and propagation of this plant. The first was an inarchment [grafting by connecting a growing branch without separating it from the parent stock] on the yellow Indian [sic] Jessamin [possibly the yellow Italian Jasmine, *Jasminum humile* L., native of the Sino-Himalayan region], and likewise on the Arabian and Brasile white Jessamin [*Jasminum sambac* (L.) Aiton, native of Asia, and *Jasminum fluminense* Vellozo, described from Brazil although earlier introduced there from Africa by the Portuguese: it has long been misidentified as *Jasminum azoricum* L. a related species endemic to the Azores, referred to by Richard Bradley in this *Dictionarium botanicum* Vol. 1 (1728) as 'the Brasil or Azores white Jasmine'];[2] but all three without succes, tho' repeated each of them twice over. The next was to make an incision at a joint, and to lay four branches down in the same soil in which the trees were planted – This succeeded very well, for in about six weeks time they were all well rooted, and fit to take off. Afterwards I took eight or nine cuttings, at different times, and with a great deal of care I got five of them to take root very well. It would have been easy for me to multiply these experiments, but as I was now master of

(1) Regarding Douglas's work as a botanist *See* Brock (C. H.), *J. Soc. Bibl. nat. Hist.*, **9**, 137–145 (1979).

(2) Advice regarding these species of *Jasminum* has been supplied by Mr Peter Green of the Royal Botanic Gardens, Kew.

twelve trees, I thought it unnecessary to give myself any farther trouble, at that time, and I have not had so good an opportunity since.

In raising these trees I used to give them frequent waterings, tho' but little at a time, and the water being always well temper'd by standing a day or two in the stove beforehand: this last I was particularly cautious about during the winter.

I found by experiment likewise, that this tree ought by no means to have the ends of the shoots cut or shortened; all the pruning it will bear is to have its lowermost branches lopt off close to the trunk.

Another caution necessary to be observ'd, is to wash the leaves often; for by long standing in the house they contract a dust, and besides are very subject to a particular sort of insect that soils them, and prejudices the young shoots, which generally lying on the under side of the leaf, may have done a great deal of mischief before they are observ'd, except we be apprised of them beforehand. They seem peculiar to the coffee leaves, for I never found them on any other plant.'(1)

Many years later when Knowlton was gardener at the Earl of Burlington's estate at Londesborough in Yorkshire he had 'grat plenty of young coffee trees' and so successful was he in their cultivation, that in 1742 he was able to send his master in London 'a small parcel of coffee', presumably prepared from these trees.(2)

It will be remembered that in February 1723 Knowlton, while working for James Sherard, had been in Holland. In May 1726 he went again, this time for Thomas Fairchild who had a nursery garden in the hamlet of Hoxton, in the parish of St Leonard, Shoreditch, then in the county of Middlesex. On his return he brought some seeds with him.(3) It is not known when Knowlton first became acquainted with Fairchild but possibly it was about 1722 as in that year he noted that Fairchild was the first in England to cultivate *Prunus lusitanica* L., Portugal Laurel.(4) A gardener to be successful in his career was advised to acquire experience in all branches of the art, including nursery gardening, and Knowlton may at some time have worked in the Hoxton garden.

(1) According to Dr V. F. Eastop, *pers. com.*, the only insect of this kind recorded on Coffee was described as a Psyllid, *Psylla coffeae*, by Vollat in *Mém. Acad. Sci. Dijon*, 1836, 228–29 (1836) and in *Compt. Rend. Séances Sci. Paris,* **3**, 72–73 (1836). It was described apparently from specimens on Coffee in a glasshouse in Paris that perhaps originated from Réunion; it has not been recorded since and its identity is uncertain.

(2) *See* below, p. 194.

(3) *See* below, p. 85. Letter to Patrick Blair (1728).

(4) *See* Aiton (W.), *Hortus kewensis,* etc., **2**, 163 (1789).

Thomas Fairchild (1667?–1729) was one of the principal nurserymen and florists of his day. He was noted for his collection of North American exotics and was considered by the botanist Patrick Blair to have the biggest collection of Aloes and succulent plants in the country.[1] He was interested in fruit culture, and in the growing of vines, and was outstanding for the many experiments he carried out for the improvement of all sorts of fruits, flowers, and trees.

Fairchild was also known for his success in growing the Guernsey Lily, now called *Nerine sarniensis* (L.) Herbert. Regarding his skill in gardening Dr James Douglas, the London physician noted earlier for his interest in botanical pursuits, wrote in his work entitled *Lilium sarniense: or, a description of the Guernsay-lilly,* 27 (1725):

> 'Mr. Fairchild has great numbers of the lillies in his gardens, in full perfection... he is so deservedly famous for his great skill in all the parts of gardening, that to him we are obliged for most of the considerable improvements that have been made in that delightful art for these several years past...'

The first known description of the Guernsey Lily was by Jacob Cornut who had seen it growing in Paris in 1634 in a garden belonging to Jean Morin, famous for its rare plants.[2] Subsequently, it was grown and written about by various authors in England. In 1680, Robert Morison, Professor of Botany at Oxford, gave the first account of how the bulbs were introduced into Guernsey in the second part of his *Plantarum historia universalis oxoniensis.* This account was in Latin, and Douglas gave an English translation:[3]

> 'A Dutch or English ship, 'tis uncertain which, coming from Japan, with some of the roots of this flower on board, was cast away upon the Island of Guernsay; the roots were thrown upon a sandy shore, and so by the force of the winds and waves, were soon buried in sand: there they remain'd for some years and afterwards to the great surprize and admiration of the inhabitants, the flowers appear'd in all their pomp and beauty...'

This story has been repeated by the island's inhabitants many times, even down to the present day.

(1) Blair (P.), *Botanick essays,* etc., 205 (1720).
(2) *See* Frost (L. C.), The history of the Guernsey Lily, *Gard. Chron.,* **154**, 430–31 (1963).
(3) *See Lilium sarniense,* etc., 2–3 (1725).

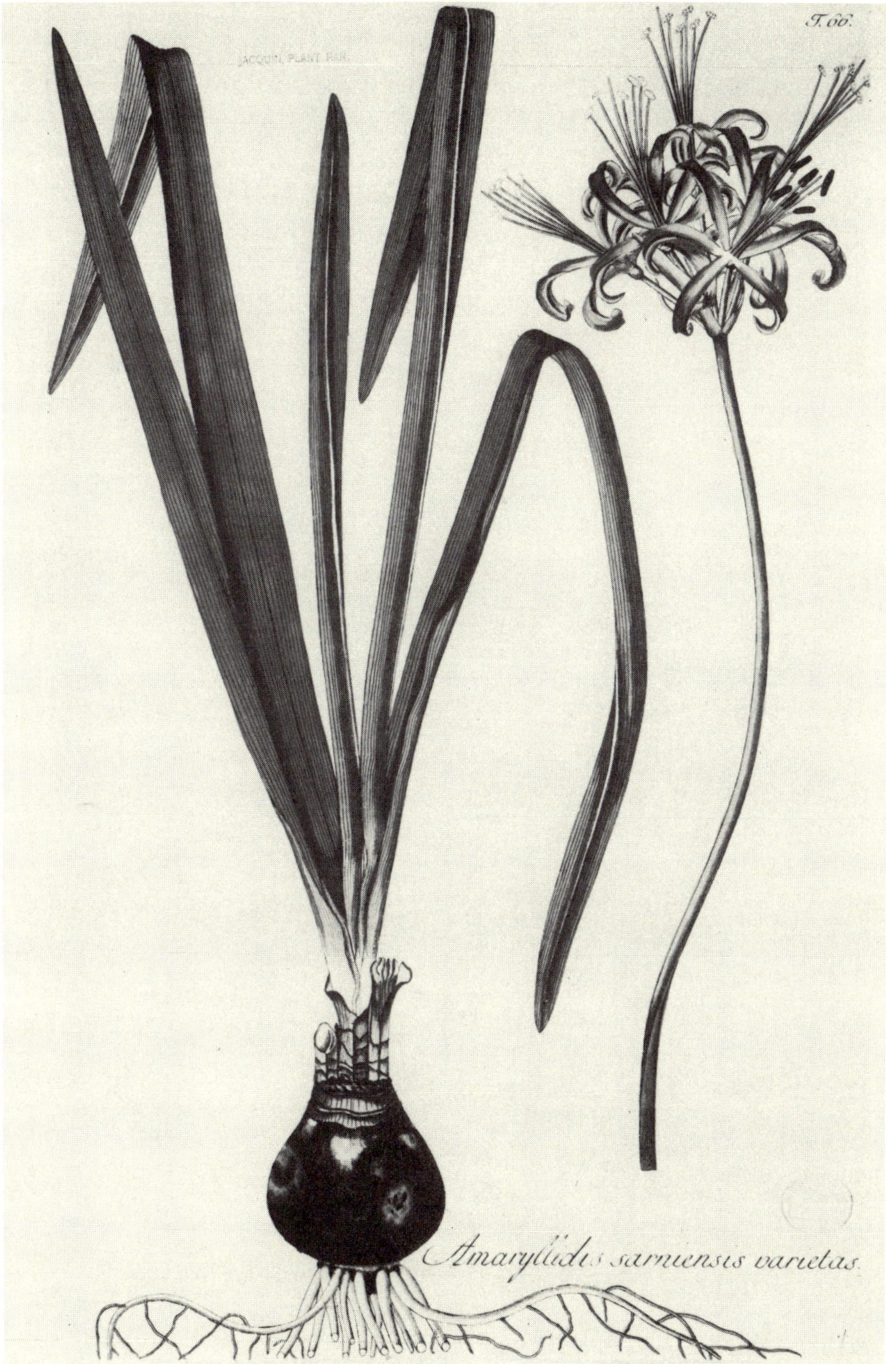

Guernsey Lily (Nerine sarniensis (L.) *Herbert, whose cultivation and history Knowlton went to Guernsey to study in 1726 at the request of Thomas Fairchild.*

Thomas Fairchild (1667?–1729), who had a very successful nursery garden at Hoxton in Middlesex. He commissioned Knowlton to visit Holland in 1726.

After publishing his book Douglas wished to extend his researches concerning the Guernsey Lily. He was anxious to obtain further details regarding its introduction to the island as well as more information on its method of cultivation there. In order to obtain the information he required, Douglas chose Thomas Knowlton, already known for his skill in horticulture as well as his enthusiasm for botany, to visit Guernsey. From 29 May 1726, when he set out in the Southampton stage coach, until his return to England about three weeks later Knowlton kept a diary.(1) In this, after giving an account of his journey, he recorded details of the people he met and of the plants they grew in their gardens. He took particular notice of the various versions told him with regard to the first arrival of the Guernsey Lily on the island, and the method used by the inhabitants in cultivating it. He noted such things as the mildness of the weather, the prevailing winds, and the type of soil. Moreover, this diary contains the names of some of the wild plants growing on the island of which McClintock states(2) that there are about a dozen first records. As well as a diary, Knowlton kept a detailed account of his expenses and after his return to England he wrote a letter to Douglas in which he gave further observations concerning the Guernsey Lily and its method of cultivation.(3)

According to his diary Knowlton set out on his journey to Guernsey at midnight on 29 May in the Southampton stage coach. The following day he had dinner at 'Fernham' (Farnham, Hampshire) and supper at Southampton.(4) On finding that he had to wait until the following Thursday before he could make a passage in a boat to Guernsey, Knowlton spent the time in making enquiries about the Guernsey Lily. However, he found no one was able to answer any of his questions until he met a 'Mr St Deneys a franch minster... a very Curious Gentleman & a grate Lover & observer in gardening.' The minister told him that he had lived many years both in Jersey and Guernsey and that in Guernsey no more care was taken in growing 'ye Bell Gurnsey, as ye Lilly there is called,' than of other common flowers such as Daffodils. Furthermore, a gentleman of his acquaintance on the island had sown some Orange seeds in the open ground two years previously, and due to the nature of the soil and mildness of the climate the seeds germinated and produced flourishing plants which he himself had seen that spring.

On 2 June Knowlton 'went abord Capt. Nathan belonging to ye Lobster boat att hamble a little town 3 or 4 miles S.E. from Southampton wth hopes to saild.' However, due to various vicissitudes, including the boat running

(1) A transcription of the diary, edited by Rosemary de Sausmarez, was published in the *Report and Transactions of the Société Guernésiaise,* **19**, 577–86 (1976).

(2) McClintock (D.), *The wild flowers of Guernsey* (1975).

(3) These three manuscripts, the diary, expense account, and letter to Douglas, are all preserved with the Douglas papers, in the Hunterian Library in the University of Glasgow.

(4) *See* the expense account.

aground and subsequently being becalmed by fog, he did not reach Guernsey until 8 o'clock on the night of Sunday 5 June. On reaching the island the gardener observed that it had 'a very pleasant appearance from [the] sea leying under a hill & y^{t} adorned and cover'd by fruit trees... y^{t} renders it one of ye Pleasants places perhaps one ever see.' The town called St Peter Port had but one church and here prayers were said every day of the week at 9 o'clock in French, and at 11 o'clock in English.

On Monday 6 June, Knowlton called on Mr Currey (Carey) and Mr Bowden in order to see 'La Belle Garnseys', but these gentlemen were busy and unable to spare time to attend to him, so: 'I walked', he wrote, 'out South Est into ye country to visit Rockes and mountainus places wch by ye sea shore are very many hear'. At Fermain Bay, about three miles from St Peter Port, he found a large number of plants that were known to him, such as:

'Samphir [*Crithmum maritimum* L., Rock Samphire]. Seabeates [*Beta vulgaris* L., ssp. *maritima* (L.) Arcangeli, Sea Beet]. Kaly Geneculatum [probably the plant called by Caspar Bauhin in his *Pinax* 'Kali geniculatum majus annuum', by Linnaeus *Salicornia herbacea,* and now included in *Salicornia europaea* L., Common Glasswort] & ye Little Sea Geranū Althea fol: & of ye Synopsis [by Ray, 356 (1754), where it has the phrase-name 'Geranium pusillum supinum maritimum, Althaeae aut Betonicae folio nostras'. It is now named *Erodium maritimum* (L.) L'Hér., Sea Storksbill].'

In two or three places on the island, one of which was near where Knowlton found the above plants, 'but always upon ye highest part of ye rockes', the gardener found some little 'bulbose roots & seeds' which he took to be 'bulbosa juncijfolia', that is *Romulea columnae* Sebast. & Mauri, Sand Crocus. McClintock points out that this is the earliest record of the plant in the British Isles.(1) The same authority writes: 'This delightful speciality of the Channel Islands bespangles the cliffs of the south and the sandy turf of the north on sunny spring days'.

Knowlton continued to walk along the shore, or near to it, for 'about fifteen miles' before turning back. He returned by an inland route and passed more houses than he had on the outward journey. The inhabitants seemed to him to be very poor, and the gentlemen almost all lived in St Peter Port as they did not have country houses, only little gardens just out

(1) *The wild flowers of Guernsey,* 235 (1975).

of the town where they went with friends to drink a glass of wine at their leisure.

On the morning of 7 June, Knowlton was invited to see the garden of Mr Elias Dubree (Dobree) the younger. Here he saw two large beds about ten feet long and three feet broad planted with Guernsey Lilies, consisting, in Knowlton's opinion, of 'about 5 or 6000 roots', that is bulbs. Next he went to see one Mr Smith, a joiner in this place who had had 'a grate many of y^{es} roots', and from there he visited Dr Bouvire (Beauvoir) at his house near 'windmill hill, about a mile and a half out of town, being full west from thence'. Here he was told the following story. A man with a parcel of Guernsey Lily bulbs had come ashore from a ship from Japan that had been either wrecked, or obliged to put in for some reason at Guernsey. The man was well received by one of the inhabitants and in gratitude he gave the islander the bulbs. These were planted but when after three or four years no flowers had appeared, the bulbs were dug up and thrown away in a corner of the garden, where, wrote Knowlton, 'in a very little time they appeared in ye most amazing beauty but how trow [true] I Leave to y^{r} more nicer judgement'. Knowlton's host gave him cultural details concerning the Guernsey Lily, and at the same time pointed out that they did not succeed in a wet soil, which was something the gardener had recognized himself. In his garden he saw Fig trees, *Ficus carica* L., many as large as Apple trees. Likewise Myrtles, *Myrtus communis* L., growing everywhere in the common ground.[1]

In the afternoon Knowlton called on Mr Bowden who took him to see his garden where again he was surprised how successfully Myrtles grew on the island. He subsequently wrote in his diary of these shrubs:

> 'Some are very Large haveing stood for many years with out sufring by frost wch is a signe of ye temperature of this Island so y^{t} I do not wonder to see... Lillys to florish so well... as they do.'

He also saw two large beds of Guernsey Lilies but in spite of the plants multiplying 'prodigiously' the owner told Knowlton that he had never observed the flowers 'to bear seed or any thing Like a seed vessell'. Mr Bowden apparently did not realize that the Guernsey Lily could propagate itself by means of offsets.

(1) According to McClintock: 'The myrtle whose abundance astonished Knowlton in 1726 will have been the common one, *Myrtus communis*.' See *The wild flowers of Guernsey*, 127 (1975).

On 8 June, Knowlton walked about a mile to another garden where he again observed great quantities of Myrtles and also large Fig trees, which confirmed his opinion concerning the mildness of the winters. However, he saw but very few Guernsey Lilies, the owner being more interested in practising the art of grafting than in growing bulbs. In one part of the garden he noticed 'a Litle mount rais'd by art about 20 feet perpendicular'. On ascending he reached an arbour with seats from which a commanding view was obtained 'of ye sea & all ye Islands &c as for example ye Gasscoins [the Casquets]'. After recording the distances in miles of the various islands to be seen from this spot Knowlton continued:

'From this place went to Mr Peater Carrys [Peter Carey] a grate Brewer who has one of ye neates Gardens in ye town [St Peter Port] or near it it being about one Litle mile from his house where he has ye Larges stock of Bell Garsey's wch are every one out of ye grown & all ye grate roots [i.e. clumps of bulbs] picked out to send to his fraInds att London ... att a modist computation thire could not be Less y^{n} 20000 roots of y^{m} being enought to Laod any Dung cart in ye Kingdon. Likewise where ye plants grow it was in com̄an kitchen Garden groun[d] (wch showes they are not so dificult in soyl ...) & y^{t} under Dwarfe pear trees wch spread & shaid y^{m} very much.'

From Peter Carey the gardener went to 'Mr Loyle' [de Lisle], a gentleman in town who had two large beds of Guernsey Lilies in a court before his house. He recorded that they had been planted:

'... 7 or 8 years & are now so thick y^{t} smal piples [pebbles] in ye street cannot be peitched together more thick y^{n} they, so prodigiously do they incress with y^{m} that have y^{m} y^{t} every year more or less they dig up one bit of a bead to send [bulbs] to thire freinds in London or other parts ... from thence [Knowlton continued] walked out to see ye Ive Castle wch Leys about one mile & halfe [from the town] being a Large peace of Grown Inclosed by a wall & within y^{t} a moate sunk to raise a mount within it wch is again Inclosd by another wall with towers att proper distances wch no dought was formaly a stronge plaice but now nothing but a heape of run's & walls cover[d] with Ive [Ivy, *Hedera helix* L.] wch I supose gives it ye name & from ye root of ye Ive heare Grow's orabranca or Rape [*Orobanche hederae* Duby, Ivy Broomrape] wch is a very prity Plante as com̄on in many plaices of England.'

McClintock in his *Wild flowers of Guernsey*, 182(1975) has pointed out that Knowlton's record of Ivy Broomrape is the earliest for the British Isles.

On the morning of 9 June, Knowlton walked 'to... Mr Marchent, another Gentleman in town'. In this garden he saw Myrtles, very large Fig trees, and also Guernsey Lilies. Here, however, there were not so many of the bulbs as he had seen in other places. Mr Marchent told him that these were sometimes called Indian flowers on account of the stories told about them. He too pointed out that although these plants flourished and increased on the island they never produced seed.

After leaving Mr Marchent the gardener went to have dinner with 'Necolas Carry' (Nicolas Carey), and subsequently he was taken to see his host's garden which was a little distance from the house. He was surprised to see 'sumer sweer majorum', *Origanum majorana* L., that had survived three winters, for in England this perennial plant has to be treated like an annual because it rarely survives our winters. In this garden he saw a bed or two of Guernsey Lilies that had been planted many years and he was informed that, at times, owing to the plants becoming so overcrowded that they choked each other, it was necessary to re-plant them. After this was done there was 'no occasion to medle with y^{m} for 7 or 8 yeare'. The surplus 'roots' were sent to friends in England.

On the morning of 10 June, Knowlton set off to visit Mr Elias Mesure (Le Mesurier), a gentleman who had a pleasant little garden that contained a great number of Guernsey Lilies which increased 'most prodigiously'. Knowlton recorded that one of the reasons these plants were successful in Guernsey was the fact that 'all these Garden[s] Ley under a hill & one [on] a declynation wch is to ye south e[ast]'. This was the very best position for any plant because the winds in Guernsey that caused damage in gardens were those from the north west and south west.

After he left this place Knowlton walked out to the salt ponds which lay about two miles out of the town in a north-westerly direction. On his way he noticed Fig trees growing in two or three places in great quantity in the hedges, but whether originally wild or not he could not tell. Among the rocks, near the salt ponds he found a number of plants including:

'Limonium [probably *Limonium vulgare* Miller, Sea Lavender] Thrift, [*Armeria maritima* (Miller) Willd.] Kaly Geniculatum☿. [probably *Salicornia europaea* L., Common Glasswort] in very grate plent[y] as Kaly spinosum [*Salsola kali* L., Saltwort]... Glaux maritima [*Glaux maritima* L., Sea Milkwort] and Erica maritima [*Frankenia laevis* L, Sea Heath].'

The manner in which the salt ponds were made is described by Knowlton as follows:

'...they first Inclose a Large Square peace of grownd with a Leavl Bottom & so dispos'd y^{t} sea watter may be Lett in at pleasure ye Grownd is Divided into small Square ponds...about 6 yards square beaded [bedded] round with pure soft clay & ye Bottom of ye same maid smoth...after ye maner our Plumer do ye sand before they run ye Lead in it w^{n} thats done they Lett in ye watter to about 3 or more inches... wch remains till ye sun Exausts it & w^{t} remains turns into salt.'

On leaving the salt ponds Knowlton called on Mr Beavior (Beauvoir) 'att ye La varange [La Vrangue]' and after dining the two men walked into the garden where Knowlton saw 'ye Double yallow rose & many other Garden plants as Myrtle's in abundance & very Large.' He observed two Orange trees planted about two or three years before, still in the best of health, and having suffered no damage of any kind during the previous winter. Under the shade of large Fig trees the men sat and drank a bottle of wine. Knowlton noted that Water-melons, *Citrullus lanatus* (Thunb.) Matsum. & Nakai, were grown in this garden 'with out any care as one well may Imagean'.

'Like wise a very Large Quantity of Lillys not less y^{n} 20 or 250000 roots [i.e. bulbs] being many Beads of y^{m} where they have stood for many years...neare 10 or 11 years since first being planted so y^{t} they cover ye whole...intirly over & perfectly shove one another out of ye Grownd being so crow[d]ed & not with standing this they blow very well...13 or 14 flowers apone one stalk & 7 or 8 Bunches of flowers one [on] one clumpe I mean by y^{t} one root with all his Incress [i.e. one bulb and its offsets].'

As to when and how this plant first arrived on Guernsey there are various stories. Mr Beauvoir's mother who at the time of Knowlton's visit was still alive, remembered that the plant was as common on the island sixty years earlier as it was then, in 1726. Mr Beauvoir himself believed its first coming was 'about 150 or 200 years ago... by a pressant to... ye Baly [bailiff?] from a master of a vessell from India', which was either shipwrecked on the coast or needed to call in at the island 'for provision or other nassarys [necessaries] for a return for favours rec'd,' the master made a gift of some bulbs. These were put into the garden and in a few years, when they flowered, these plants recommended themselves by their beauty to all who saw them. All the many plants of the Guernsey Lily to be seen when Knowlton paid his visit were considered to have 'sprung from those few roots, 'said Mr Beauvoir 'for thires non think it to be a native of ye Island'.

On 11 June, Knowlton went to see Mr Measurir (Le Mesurier) a minister in charge of 'ye Collige or free Scholl for ye Islanders' who confessed that he always had the opinion that the Guernsey Lily was too common a plant to take much notice of. However, he was soon persuaded to think otherwise when Knowlton 'Desired him only to observe ye root from ye time of flower till ye next spring &c w^n^ it has ye green Leaves one [on]'.

Mr Le Mesurier told Knowlton about a tree in a neighbour's garden that they called:

> 'Balsame De prow [Peru] w^ch^ [said the gardener] I went to see & take to be ye tachamahca of Pluknett[1]... it... was brought from St Malos & first from ye west Indies where it florishes in ye natural grown & is a Large fine tree smel's very Balsamick affording a clamy substance att ye end of ye Shoo-tts or buds they had only one tree of it otherwise I had brought one but have ye promise of a plant of it w^n^ ye Season comes for I showed ye gen-tleman how to Lay it.'

Regarding this tree it is of interest to note that in the Sloane herbarium, Vol. H. S. 318. f. 38, preserved in the British Museum (Natural History), there is a specimen of a twig, 'Sent by Mr Parey from Guernsey as the leaves of the Tree of ye balsam of Peru, growing in that Island.' Little appears to be known of Mr Parey except that he wrote two letters to Sir Hans Sloane from Guernsey, dated 1734 and 1735 respectively, on the Balsam of Peru. In the first of these he enclosed a small twig from the tree in Guernsey. In 1973, McClintock published a paper in the *Quarterly Journal of Forestry*, Vol. **67**, 136–140, in which he pointed out that the true Balsam of Peru comes from the leguminous shrub *Myroxylon balsamum* (L.) Harms var. *pereirae* (Royle) Baillon, from Central America, but that the twig Parey sent Sloane and which is now preserved in the Sloane herbarium is in reality a specimen of *Populus tacamahaca* Miller, Balsam Poplar. The same authority has also drawn attention to the fact,[2] that this specimen is the earliest known of

(1) Plukenet (L.), *Phytographia*, Part 3, tab. 228, fig. 2 (1692).
(2) In *The wild flowers of Guernsey*, 153 (1975).

any plant from Guernsey. Knowlton's plant was therefore *Populus tacamahaca*, now called *Populus balsamifera* L.

Knowlton always took a keen interest in botanical and horticultural books, especially those with illustrations, and his quick recognition of this tree in the garden in Guernsey is an example of his skill in the identification of plants. Like many of his contemporaries he made much use of Leonard Plukenet's work of 1691–1705 with its many plates, each containing a number of small figures of plants.

In Mr Le Mesurier's garden Knowlton saw an Orange tree which was growing in the open ground which flourished and looked very well, and here again he saw many Guernsey Lilies. After this he walked out into the country 'about six miles' to a parish called St Andrews, west-south-west of St Peter Port. His intention was to visit a minister's garden but the owner was away from home so he could not have the pleasure of seeing it. He had been informed that a great many Guernsey Lilies grew there. However, a friend of Knowlton's, a Mr Martian (Martin), later told him that he had been to the garden and seen very few of the bulbs. Furthermore, he learnt from his friend that in Guernsey it was customary for gentlemen to live in town, that is St Peter Port, and to have 'a little garden just without'.

Finally, on 12 June, Knowlton went to see the garden of 'Mr Jn Dubrea' (John Dobree) who had 'ye gratest Quantity of ye roots in all this Island'. Here the gardener appears to have had some difficulty in understanding what was told him, complaining that, as the Islanders 'all speake thir own Language wch is Brooken franch & some but very Indifrant English so one may easily mistake y^m.' However, Knowlton discovered that the great quantity of Guernsey Lilies growing in Dobree's garden had originated from a 'few roots' planted by Mary Dobree in her father's garden about 22 years previously, the 'few roots' having been found by Mary lying in the 'Highway'. These details Knowlton obtained from a note written by Dobree's son-in-law.

Knowlton was back in England by the middle of June, having spent a total sum during the whole of his trip of £4.0s. 5d. The detailed account of his expenses reads as follows:

[May 29, 1726]	Sett out in ye staige coatch for Southampton	18–00
	for my Diner Fernham [at Farnham] & Bear [beer]	1–00
	for Beare by ye way.........................	00–06
	att Southampton for supper &	00–08

31	for Bread & cheas & Beare	00–06
	for Diner & Bear	00–09
	for Bear att night	00–04
June ye 1	for Beare Bread & chesse	00–06
	for Diner & Beare	00–09
	for Beare att night	00–04
2	for Bread & chesse & Beare	00–06
	for 3 nights Lodging	00–09
	for passing ye ferry to Hamble	00–01
	for Bread cheass Baccon Beare to carry one [on] Borde	03–09
	for going one [on] Borde	00–02
	being one [on] out passage from ye 3 to ye 5 & paid to ye Capt for itt	5–00
	for super ye night came into Garnsey	00–10
6	for Brackfast & Beare	00–06
	for Diner & whine	01–00
7	for Bread chese & whine	00–06
	for Diner & whine	01–00
	for Supper & whine	00–06
8	for Brackfast	00–06
	Dined att m^{r} Beavers	
	for super & whine	00–09
	for washing one shirt	00–03
9	for Brackfast	00–06
	Dined with Mr Carys	
	for supper wine	00–06
10	for Brackfast & Beare	00–06
	for Diner & wine	01–00
	for wine	00–06
11	for Diner & wine	01–00
12	for Brackfast	00–06
	for Diner	01–04
	for supper	00–06
	for a cloth & Baskett for ye Bell Garnseys...	00–06
13	for Brackfast	00–06
	for Diner & whine	01–00

14	for 9 night Lodging att Garnsey............	2–03
	for 4 qts of wine for 2 of Beare	2– 7
	for Bread Butter	00–10
17	for my passage to the 17th..................	05–00
	& gave ye men wch is a custome............	01–00
	for ye passing ye ferry att Southampton......	00–01
	for my dinner att Hamble..................	01–00
	for my Exspense att Southampton..........	01–00
	for coatch hire	18–00
	for Diner att fernham [Farnham] & Beare ..	1– 4
		£4–00– 5

In a long informative letter, dated 'June ye 30th 1726, Buckingham house', Knowlton gave Dr James Douglas details concerning his recent visit to Guernsey. He wrote as follows:

'Worth S.^{r}
As I have ye hounour to be sent over to Garnsey att y^{r} charge one [on] purpos to Learn ye culture of ye Bell Garnsey y^{t} was so agreeable to my Inclynations y^{t} I take it to be my Duty to give you my opinion of y^{m}...(& wonder...y^{t} no curious person...attemptd it before)...first how I have observed y^{m} thire & ye reson why they will not flower with us, tho ye same root will many times in y^{t} Island as is obeserved by some & Denyd by others insomuch y^{t} ye R.^{d} Mr Jn ^{o}Measures [John Le Measurier] tould me he thought y^{t} ye root Dey'd after it flowered...but y^{t} is clearly prowed [proved] to ye contrary By...ye worthy d^{r} Douglas...'

Douglas's work, published in 1725 under the title of *Lilium sarniense: or, a description of the Guernsay-lilly*, is illustrated by two plates. These contain figures of the roots, bulb, leaves, and flower, of this plant, a bulb which has produced flowers together with its offsets, and various dissections of the bulb and flower, including a dissection of a dormant plant (one which had flowered) showing the 'embryo' of the next season's growth.

The season of growth of the Guernsey Lily is from September to May and due attention therefore must be given to providing the plants with protection from winter frosts. They require a sheltered, warm, and well drained situation, and in the summer after the leaves die down the bulbs need sunshine during their period of rest to ripen. The bulbs increase by offsets, and clumps must be periodically divided and replanted to ensure regular flowering. Knowlton was aware of these requirements for the successful culture of Guernsey Lilies, a fact made clear in his letter to Douglas. Briefly, the reasons he gave regarding the non-flowering of these bulbous plants in English gardens were as follows. In Guernsey:

'...they Injoy ye most temperate winter's a people can ever wish for haveing hardly ever any frosts, tho heare [in England] to ye Gratest Degree Imaginable... ye winter being ye time of thire groweth and ye Leaves y^{n} tender & Succulent are very apt to be... Injured & pinched... our Growners in ye winter being for ye most part Gluted with raine & of a clayey cold Quality... will retain it ye more & this plant being a grate enemy to it, y^{t} even in Garnsey... it will not thrive thire in any Inclyning to a most [moist] soyl as many Gentlemen have relatied to me & Likewise from my own observation in St petters port & neare y^{t} town in Gentlemens gardens these flowers are all cultivated in such grate n.^{ers}... & most one [on] ye Declivity of Hills to ye Southeast & ye soyl generaly a sandy Loom with Depth enough & underneathe ye same is rockes y^{t} Drain's it so well y^{t} thire never stands any raine stagnating to chill & sok ye Bulbs wch is as grate an enemy as frost & these being ye only resons we cannot have y^{m} flowers as they should.'

Knowlton pointed out, that in order to protect the plants from frost and stagnant water it was necessary to make a specially prepared bed, with good drainage and suitable soil. Once the bed was made and planted the bulbs could be left undisturbed for eight to ten years. They could then be lifted in the manner following, as practised by the people of Guernsey:

'...they begin att one End of a bead & Dige up ye roots clean & so pick out ye Largest roots to send to thire frainds in England ye rest they plant as before or else throw y^{m} away as they say & this is thire meathod every year y^{t} it is Impossable to say w^{t} a grate Incress they have by this management.

If I ware to advize any person to plant... ye roots I would not have Less y^{n} 100 of y^{m} & out of such a Quantity & good management one might be shure to have several flowers every yeare & in a very Little time would have very grate no.^{rs} & will I believe do better this way y^{n} in potts therefore I would not advize y^{t} way... it being not so natural & ye earth much sooner exhausted besides more subject to heats & colds wch to prevent must be housed in ye winter...'

Of the first coming of the bulbs to Guernsey, Knowlton related:

'... ye story is a shipe coming from japon or India was ether by contrary winds or shipwrack oblidge to put into y^t Island & being very kindly entertainde by one Mr Somere [Jean de Sausmarez] Grandfather to ye presant Mr Somere [Henry de Sausmarez] Auther of ye marine Survayer... ye master gave him 6 of these roots wch he planted..., att Last a flower sprang up amungest ye weeds wch being discovered by his servant was... brought to him & in a Litle time he recolected how they ware gave him and this story was tould me by Mr Henery Somere [Henry de Sausmarez]... Likewise this was ye story he... heard his Grandfather relaite w^n he was a Boy.'

Attention may here be drawn to the fact that on his return from Guernsey Knowlton contacted Mr Henry de Sausmarez, a native of the island, in order to check on a story relating to the bulbs' first coming to Guernsey as published by Richard Bradley. This author, in the Appendix to his *New Improvements of planting and gardening... The fifth Edition*, 591 (1726), wrote that he had been informed by:

'M. de St. Marets a very curious gentleman of that place... that in his grandfather's time a ship from China came in there, when some sailors gave a parcel of these roots to a person who then kept a publick house... from thence came the whole stock that is in the island.'

When Dr Douglas published *A description of the Guernsay-lilly* in a second edition,[(1)] he recorded this version, corrected the name 'de St. Marets' to 'Saumarez' (it should, in fact be de Sausmarez), pointed out that it was Henry de Sausmarez's grandfather who received the roots and 'not one who kept a publick house', and gave a 'History of this plant's coming into Europe, and particularly into Guernsey.' This included four different accounts told him by Knowlton, concerning the bulb's first arrival on Guernsey. The fact that no documentary evidence is available confirming the truth of any of these stories has resulted in much speculation with regard to their reliability and accuracy. Various persons have considered Robert Morison's story to be a fabrication. It will be remembered that he recorded that a ship from Japan was wrecked off Guernsey, some bulbs of the Guernsey Lily were thrown upon the sandy shore, they became

(1) i.e. in the work when published in 1729 and 1737, and called on each occasion 'second edition'.

covered in sand, and after several years flowers appeared. In 1978, Dr Helen Brock, in her paper entitled 'A tale of mystery and imagination', read before the members of the Society for the Bibliography of Natural History, rejected Morison's story and considered that it had been concocted.

In 1753, in *Species plantarum*, **1**, 293, Linnaeus gave the Guernsey Lily the Latin name *Amaryllis sarniensis,* adding: 'Habitat in Japonia, nunc in Sarniae insula Angliae [Grows in Japan, and now also in the island of Guernsey, England]'. In 1821, William Herbert transferred the plant to a different genus when it was called *Nerine sarniensis.*(1) In *Flora capensis* vol. **6** it is shown to be endemic to South Africa, with a note reading: 'Commonly cultivated in European gardens under the name of Guernsey Lily.' The plant has been called by a number of other common names including, the Narcissus of Japan, Le Belle Guernzey, and the Indian flower.(2)

We must now return to Knowlton's letter. He told Dr Douglas that as the Guernsey Lily was a native of a country with a warm climate it would need protection when grown in English gardens. He wrote:

> '...as this is so beautifull & Lovely flower it will deserve our utermost care for its persevation & shure I am from 5 or 6 years practice of a Large no$^{r.}$ under my care that they Gratly suffer'd by y^{e} frosts every winter tho planted under a South wall & in a dry soyle & well cover'd with peas straw in y^{e} voylance of y^{e} winter & notwithstanding all that I could not y^{n} perserve y^{e} Leaves from being fros[ed] so much that they weare turned into a muculage [mucilage].'

Thus, in order to protect these plants from cold and frost as well as excessive wet, Knowlton invented a frame made of wood with sliding shutters which could be placed over the bed in which the Guernsey Lilies were planted. Of this he said:

> '...ye time of removing & planting y^{m} is about ye midle of June to ye end of july...about St Michellmus [29 September] it will be nassary to have a fram[e] of wood to be so maide as to Inclose y^{e}...bead, ye back part about 6 inches hight ye forepart 4 & so Banked up with Earth all round ye out

(1) *See* Herbert (W.), *An appendix* (at end of 'A treatise on Amaryllis') in *The Botanical Register,* **7**, 19 (1821).

(2) *See* Douglas (J.), *A description of the Guernsay-lilly,* the second edition (1729) and (1737).

side up to ye tope to pervent ye frost &c. & Likewise some wooden Shutters to cover ye tope with all as accasion may requir just such as ye Blomest in Holland have to cover thire curious bulbs wch may be put one [on] or off att pleasure to give y^m all ye advantaige of aire... Likewise if ye frost should be very Sharpe you ought to covere these frames with some horse Dunge... with y^{es} Dericetions I hope ye curious will have ye satisfaction to see y^m flower.'

Knowlton informed Dr Douglas that some of these bulbous plants had been presented to:

'... ye Grate professer of Botany att... Leyden in Holland[1]... by ye worthy Jn °Bauvouel Esqr [Beauvoir] of ye Lavrangue [La Vrangue] in Garnsey... Likewise some [had been given] att ye same time to ye famious Monsir La Cour of ye same city where [wrote Knowlton] I see allmost 4 or 500 of thes[e] roots this May 1726... I think it was ye only Garden y^t had any of y^m in a quantity ye Gardener tould me y^t he flower'd some of y^m every yare wch was maniaged in ye same method I perpose.'

Pieter de La Court (1664–1739) was a wealthy cloth merchant with a garden near Leyden who, in the words of the eighteenth-century botanical writer Richard Bradley, was 'a gentleman of extraordinary skill in the affair of gardening.' In the following passage Knowlton expressed his opinion concerning the state of horticulture in Holland and England, praising the Dutch but also giving a pat on the back to his late master James Sherard who had constructed greenhouses and stoves in his Kentish garden in order to cultivate exotic plants. He pointed out that, to our neglect and shame, a people as we are and counted ingenious should be so beholden 'to ye Little people [the Dutch]'[2] for first showing us how to grow such plants as Pineapples and Coffee, as well as:

'... most of our Beautyfull ornaments for stoves Green's houses &c wch of Late have thro ye Industry of a Mr [James] Sherard ben maide prity femerliare amoungst us... I Likewise take the Liberty to Lett world know y^t ye worthy gentleman Last named was ye first y^t over designed our now new

(1) Peter Hotton (1648–1709) Professor of Medicine and Botany 1695–1709. *See* Douglas (J.), *A description of the Guernsay-lilly,* the second edition, 19 (1729) and (1737).

(2) When Knowlton wrote of 'ye Little people' he may have been referring to the population of Holland which, at that time, was small compared to that of England.

Improved stoves to y^{t} perfection we now have y^{m} in tho Mr Bradly [Richard Bradley, the horticultural and botanical writer] & others have endeaverd to Robe him of y^{t} honour which is so much his dew therefore I hope thaly not blame one for relating fact when at y^{t} time I had ye honour to searve him...'

In his second edition of *A description of the Guernsay-lilly*, Douglas included under the heading 'Cultura', information collected by Knowlton concerning the plant and its method of cultivation on Guernsey. At the same time he gave details for growing the bulbs in this country and in so doing, described the frame made of wood contrived by Knowlton for protecting the bulbs in winter-time from cold, frost, and excessive rain.[1] Knowlton was quite correct when he said that the Guernsey Lily needed protection when grown in English gardens. According to Mr C. Brickell, Director of the Royal Horticultural Society's Garden at Wisley:[2]

'In Britain generally, it is not considered to be hardy out of doors and is almost always grown as a cool greenhouse plant. It will survive and flower in sheltered gardens in Cornwall, the Scilly Isles, and in the Channel Islands. In the rest of Britain, however, it has to be grown in a cool greenhouse as although it may survive in one or two areas out of doors it seldom flowers to any great extent.'

Finally, it is of interest to note that bulbs of *Nerine sarniensis* that grew so readily on Guernsey in the eighteenth century are now 'to be found only in one very sheltered garden – against a wall, on a slope, facing south up-hill'. This information has been supplied to the writer by Mrs Ashley (Miss Rosemary de Sausmarez)[3] who read Knowlton's diary of his visit to the Island to members of Le Société Guernésiaise as well as publishing a transcription in the Society's *Transactions* in 1976.

Knowlton addressed the above letter of 30 June 1726 to Dr James Douglas from 'Buckingham house'. This indicates that he may have worked there for a time before being appointed gardener at the Earl of Burlington's estate at Londesborough in Yorkshire. Buckingham House, built in 1705 for John Sheffield (1647–1721), third Earl of Mulgrave, created on 23 March 1702/3 Duke of Buckingham, was described by John

(1) *See* p. 47, etc. (1729), and (1737).
(2) In epist. 22.vi. 81.
(3) In epist. 30.ix.82.

Macky in the first volume of his *A journey through England* (1722). After referring to it as one of the great beauties of London, both by reason of its situation and its building, he wrote, p. 194:

'It is situated at the west end of St. James's Park fronting the Mall, and the great walk; and behind it is a fine garden, a noble terrace (from whence, as well as from the apartments, you have a most delicious prospect) and a little park with a pretty canal.'

The Duke died at Buckingham House on 24 February 1720/1 and bequeathed the property to his Duchess, Katherine.

It is possible that when Knowlton returned from Guernsey there was a post vacant at Buckingham House and that Dr James Douglas, and also perhaps Thomas Fairchild, recommended him for the appointment. Both men knew the gardener as a reliable workman skilled in his profession.

A description of the garden at Buckingham House is given in a letter written by the Duke and published after his death in the second volume of his *Works* (1726).(1) In this letter, after referring to the 'far distant prospect of hills and dales, and a near one of parks and gardens', the writer continued:

'To these gardens we go down from the house by seven steps, into a gravel-walk... with a covered arbour at each end of it. Another of thirty foot broad leads from the front of the house, and lies between two groves of tall lime-trees planted in several equal ranks upon a carpet of grass: the outsides of these groves are bordered with tubs of bays and orange trees.

At the end of this broad walk, you go up to a terrace four hundred paces long... On one side of this terrace, a wall covered with roses and jassemines is made low to admit the view of a meadow full of cattle just under it, (no disagreeable object in the midst of a great city) and at each end of descent into parterres, with fountains and water-works.

From the biggest of these parterres we pass into a little square garden, that has a fountain in the middle, and two green-houses on the sides... and near... lies a flower-garden. Below all this, a kitchen-garden full of the best sorts of fruit, has several walks in it fit for the coldest weather.'

(1) *See* A letter to the D[uke] of Sh[rewsbury], 218–20. Buckingham House was situated at the west end of St James's Park, fronting the Mall. It was on the site of the present Buckingham Palace which was built in the reign of George IV. *See* Wheatley (H. B.), *London past and present,* etc., **1**, 293 (1891).

Adding to the peace and beauty of the place there was also 'a little wilderness full of blackbirds and nightingales'.

At the end of 1726, Knowlton was working for the Earl of Burlington at his Yorkshire estate on the edge of the Wolds at Londesborough, in East Riding. It was about 17 miles south east of York and about the same distance from Hull. The house, Londesborough Hall, stood on high ground and it had, as described by Defoe,[(1)] 'a noble prospect towards the Humber, and also towards the Woulds.'

The exact date when Knowlton took up his new appointment is not known. The Londesborough estate accounts, now preserved at Chatsworth, (Vol. 298), record that he succeeded a gardener named Nathaniel Adams and what seems to be his first payment is entered under the date 23 December 1726, where we read:

> 'Paid Mr Knowlton Gardener being money to be applyd to the payment of the severall workmen concerned under him in the garden at Londesburgh and for other disbursements for which he be accountable as by his receipt £12 4s. 8d.'

Knowlton was very happy at Londesborough where he remained until he died, and during the years he lived in Yorkshire he corresponded with some of the most eminent naturalists of the day. Much of the rest of the present work will be devoted to extracts from these letters, but first let us consider briefly the life of the nobleman Knowlton served for so many years, and some details of interest concerning Burlington's Yorkshire estate at Londesborough.

Richard Boyle, born 25 April 1694, was a child when, after his father's death 9 February 1703/4, he succeeded as third Earl of Burlington and fourth Earl of Cork. With these titles he inherited estates in Ireland, Yorkshire, and London, and his chief residences were Lismore Castle in Ireland, Londesborough Hall in East Riding, Burlington House in Piccadilly, and Chiswick House in Middlesex.

After his father's death Lord Burlington was brought up by his mother who, until he was grown up, kept house for him at Londesborough and in London. How often during his early years he visited his Yorkshire estate is not known, but David Neave records[(2)] that from 1720 until his death in

(1) [Defoe, D.], *A tour thro' the whole island of Great Britain... By a gentleman. The second edition,* etc., **3**, 133 (1738).

(2) Neave (D.), Lord Burlington's park and gardens at Londesborough, *Garden History, Journ. Gard. Hist. Soc.,* **8**, 69 (Spring 1980).

Richard Boyle, third Earl of Burlington (1694–1753), who employed Knowlton as his gardener at Londesborough, had a great influence on both architecture and garden design and was patron to many of the writers, musicians, painters, sculptors and architects of his day.

1753 Burlington paid almost annual visits to Londesborough. However, there were long periods when Yorkshire saw nothing of him. Dr P. Roebuck in his paper entitled 'Absentee landownership in the late seventeenth and early eighteenth centuries' has given a number of reasons for landlords such as the Earl of Burlington being away from their properties.[(1)] For example, many landowners held property in more than one county, some in several:

'Moreover towards the end of the seventeenth century public services began to make growing demands on the upper reaches of landed society ... Many landowners acquired town houses in London or provincial boroughs, spending less of each year on their estates... Developments in the London money market... played a part in the growth of absenteeism. Landowners... heavily engaged in the speculative booms which succeeded one another from the 1660's... preferred for obvious reasons to direct their activities from bases in the metropolis... Social developments... drew landowners away for longer periods from their estates... from the end of the seventeenth century 'the grand tour which followed University became almost *de rigueur*' for the eldest sons of wealthy landowners. Thereby they acquired an early taste for travel which they frequently later indulged...'

When the spirit of building pervaded the opening decades of the eighteenth century new residences and streets were being built in London in the vicinity of St James's Park, and not far distant from this fashionable new area the Earl of Burlington's property north of Piccadilly occupied an extremely valuable site. It was not long before the Earl released some of his land for development, subsequently taking an active part in controlling and supervising the work done. Moreover, as we shall shortly see, in 1716 he began to make alterations to Burlington House.

Burlington had not yet reached the age of 21 when, in May 1714, he set off on a grand tour of Belgium, Germany, Switzerland, Italy, and France, and on this journey he purchased a number of pictures and other works of art. His interest in Palladian architecture is considered to have been first aroused with the publication in 1715 of the first volume of Colen Campbell's *Vitruvius britannicus,* and Giacomo Leoni's edition of Andreas Palladio's work on architecture which also began to be issued in 1715.

In the year that alterations were commenced at Burlington House, James Gibbs (1682–1754) designed the east side of the forecourt.

(1) *The Agricultural History Review,* **21**, 11–14 (1973).

Subsequently in 1719, after the Earl had again visited Italy when he spent several months in Vicenza studying Palladio's work, he employed Colen Campbell (1676–1729) to remodel Burlington House in the Palladian style. Here several ceilings were later painted by William Kent (1685?–1748). Burlington brought Kent back to England with him in 1719 having first met the artist in Rome in 1714/15. He intended to employ him as a history painter but it was as an architect and landscape gardener that Kent showed his true talents. His landscape gardening displayed the new taste for garden design which marked a departure from the formal style. He viewed garden landscape through the eyes of a painter and the element he introduced when he assisted Burlington during the third decade of the eighteenth century in planning the garden at Chiswick House was the picturesque, or, in other words, the pictorial.

When, in 1725, Chiswick House was partly destroyed by fire, Burlington began a new villa to be connected with the old house by a link building. This project the Earl had been considering ever since his return from Italy in 1719. It was modelled partly on Palladio's masterpiece the Villa Capra (or Rotunda) outside Vicenza, and on other buildings including some villas by Palladio's pupil Vincenzio Scamozzi, and was to serve as a repository for Lord Burlington's ever increasing number of books, pictures and other works of art. It was also to be a place in which the Earl could entertain his many friends and visitors from home and abroad. Lees-Milne, in his *Earls of creation*, 154 (1962) wrote:

> 'Chiswick was Lord Burlington's favorite retreat, where he most loved to be and where he never tired of making some improvement or other... He was even known to dash down from Yorkshire to spend ten days at Chriswick, and back again.'

Burlington appears to have become seriously interested in architecture when he was about 22 years old. During his second visit abroad he was so impressed with the buildings by Palladio and his colleagues that he became determined to reinstate the principles of Roman ('regular') architecture in England – a style practised here a century earlier by the architect Inigo Jones (1573–1652). Burlington played an important part in the revival of neo-classicism in England and a list of buildings with which he was connected, as a practising architect, is published in Colvin's *Bibliographical dictionary of British architects 1660–1840*. He not only offered support and encouragement to architects and sculptors, including Colen Campbell, G.

B. Guelfi and John Michael Rysbrack, but he also gave his patronage to writers, musicians and painters, among whom were Alexander Pope, Jonathan Swift, John Gay, George Frederick Handel, and William Kent.

On 21 March 1720/21, the Earl of Burlington was married to Dorothy, daughter of William Savile, second Marquess of Halifax. Of their union there were three children, all daughters, but only one, Charlotte Boyle (1731–1754) survived Burlington. She married William Cavendish, Marquess of Hartington (1720–64), later Duke of Devonshire, and when the Earl died in 1753, Charlotte as his sole heir brought to her husband's family all her father's estates. However, she died the following year in 1754; her father-in-law died in 1755, and her husband then succeeded as fourth Duke. Thus to the immense Devonshire estates which included Chatsworth, the famous mansion on the banks of the River Derwent in Derbyshire, were added Burlington House, Chiswick House, and Londesborough Hall in Yorkshire. After Burlington died, Thomas Knowlton remained at Londesborough in the service of the Duke of Devonshire.

While Chiswick was Lord Burlington's favourite retreat, Londesborough claimed a good deal of his attention. In fact, as pointed out by Neave:(1)

> 'Burlington . . . throughout his adult life spent part of almost every summer or autumn away from the pleasures of London and Chiswick in his distant rural retreat of Londesborough. The visits varied in length from two or three weeks to the most lengthy of seven months in 1738–9.'

The journey from London which must have been far from comfortable took three days. To cross the river Ouse Burlington could either take the ferry east of Goole or go a far longer way round by road and cross the river by bridge at Selby. In bad weather it took three and a half hours to be rowed across the river, and in addition to the discomfort of the journey there was always the possibility of being robbed by highwaymen along lonely stretches of road.(2)

The Earl of Burlington early in life became involved in public affairs. He was Vice-Admiral of the county of York 1715–33, and Lord-Lieutenant of the East Riding 1715–21, and of the West Riding 1715–33. In 1722, he was elected a Fellow of the Royal Society and in 1724, a Fellow of the Society of Antiquaries. In 1730, he became a Knight of the Garter. Towards the end

(1) Neave (D.), *Londesborough, history of an East Yorkshire estate village,* 15 (1977).

(2) For details of these journeys *See* Lees-Milne (J.), *Earls of creation,* etc., 161–62 (1962).

of his life Burlington suffered from illness, and from Knowlton's letters it is learnt that in March 1750/51 he visited Bath, a city noted for its medicinal waters, after which his health appears to have improved. However, this improvement was only temporary and on 3 December 1753 the Earl died at Chiswick, having only returned from Londesborough at the beginning of the previous September. His body was taken to the church at Londesborough and he was buried in the family vault there.

Not long before Burlington inherited Londesborough a bird's eye view of the house, surrounding grounds, and village, was drawn by Leonard Knyff (1650–1721), a Dutchman who came to England about 1690 and settled in Westminster. The drawing was engraved by Knyff's fellow countryman and neighbour at Westminster, Johannes Kip (1653–1722), about 1700.[(1)] The H-shaped house depicted in this engraving (p. 73) was built during the sixteenth and seventeenth centuries. The central, and oldest part, facing east and west, was built, it is said, in 1589, while the two wings at the north and south ends of this central block were added in about 1680. During the years Lord Burlington was owner of Londesborough there is no record of his ever carrying out any major work to the exterior of the house. With regard to the park and gardens, however, the story was very different.[(2)]

Kip's view shows a flight of steps leading from the terrace opposite the south front of the house to a semi-circle of trees from which ran a wide avenue going south into the park, where are depicted several deer. It is known that in 1724 there were 600 of these animals, and that the park consisted of about 229 acres. Gradually Burlington increased this area until it finally covered some 700 acres. Meadow and agricultural land was converted into parkland, and much of the hamlet of Easthorpe was demolished for the same purpose.

Londesborough Hall originally stood on raised ground which fell sharply away to the south-east. In order to provide a level site for a formal garden the ground to the east of the building was raised behind a terrace wall, and a sunken lawn was then made with, at the eastern end, a circular pond with a fountain. This area was apparently known as 'the fountain

(1) *See* plate 31, undated but probably engraved about 1700, in *Nouveau théatre de la Grande Bretagne,* etc. (1708) London.

(2) In writing of the house and grounds at Londesborough, the Compiler is deeply indebted to David Neave for his works on the subject, namely, *Londesborough, history of an East Yorkshire estate village* (1977), published by Londesborough Silver Jubilee Committee, and two articles entitled 'Lord Burlington's park and gardens at Londesborough, Yorkshire', *Garden History, Journ. Gard. Hist. Soc.,* **8**, 69–90 (Spring 1980), and 'Londesborough Hall', *The Georgian Society for East Yorkshire* No. 5 (1978).

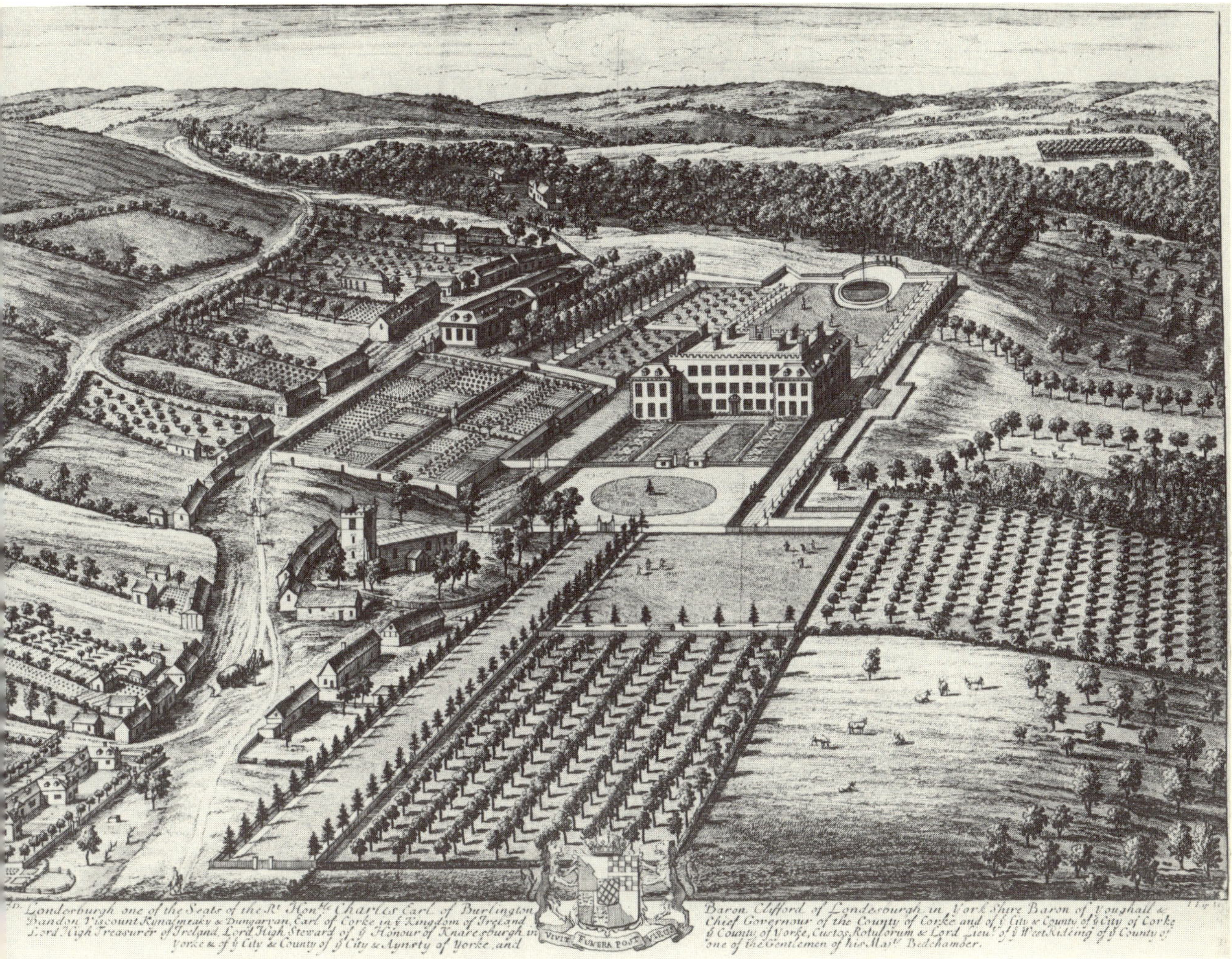

Londesborough, a bird's eye view engraved by Johannes Kip from a drawing by Leonard Knyff, c. *1700.*

garden'. During the third and fourth decades of the eighteenth century, when various alterations were being carried out, this garden was left undisturbed. So was an avenue, probably of Yew trees [*Taxus baccata* L.] and later known as Garrick's Walk, to the west of the hall. Before changes were made this avenue had a bowling green and a plantation on one side and some houses and Londesborough church on the other. Burlington redesigned the pleasure garden to the west of the house, and in about 1730, he removed the walled kitchen garden from its position north of the hall to a new site of four acres in the valley to the south-west adjoining a water-mill.

After Knowlton replaced Nathaniel Adams as gardener at Londesborough considerable sums of money were spent on the purchase of trees, plants, and seeds, and the increase in expenditure on wages is evidence that more labourers were taken on in the park and gardens. During the years 1728–32 Estate Accounts show that many hundreds of trees were purchased. For example, four entries in these accounts read:

20 June 1728 Paid Nicholas Wight in full for Ash [*Fraxinus excelsior* L.]. Elm and other trees £14= 5=0
22 September 1730 Paid Nicholas Wight in full for 300 large English Elms [*Ulmus procera* Salisb.] £15= 0=0
24 March 1731 Paid Thomas Jackson for 200 Scotch Firrs [*Pinus sylvestris* L.]........................ £7=15=0
11 November 1732 Paid John French for 400 Spruce Firrs & 4000 Scotch Firrs [*Pinus sylvestris* L.] ... £8= 0=0

It is not possible to be certain which species of Spruce the nurseryman John French sent Knowlton but it was probably *Picea abies* (L.) Karsten, the Norway or Common Spruce. It was in cultivation in Britain by the seventeenth century, and is an important forestry tree. It grows well on moderately fertile moist soils, providing they are not liable to dry out. Knowlton also purchased trees and plants from the Yorkshire nurseryman, John Telford of York.

In 1739, an estate plan of Londesborough was made by Thomas Pattison which shows the changes that Burlington carried out up to that date in the park and gardens (p. 75). The transformation of the pleasure gardens to the west of the house is best told by Neave who, after studying Pattison's plan and a poem entitled 'Lanesborough Park' by Mr Wyld of Otley, published at York in 1738, wrote:(1)

'The yew avenue [i.e. Garrick's Walk] was retained but to the north a wilderness area with winding walks, terminating in small open spaces with features such as vases or statues, was created by the demolition of the houses and cottages to the west of the church and the extension of the gardens to the village street. South of the walk the bowling green was replaced partly by a rectangular pond with apsidal ends from which ran a formal walk westwards which expanded in stages to form a large open area then contracted to a walk again and ended in a circular open space...'

(1) *See Journ. Gard. Hist. Soc.*, **8**, 77 (Spring 1980).

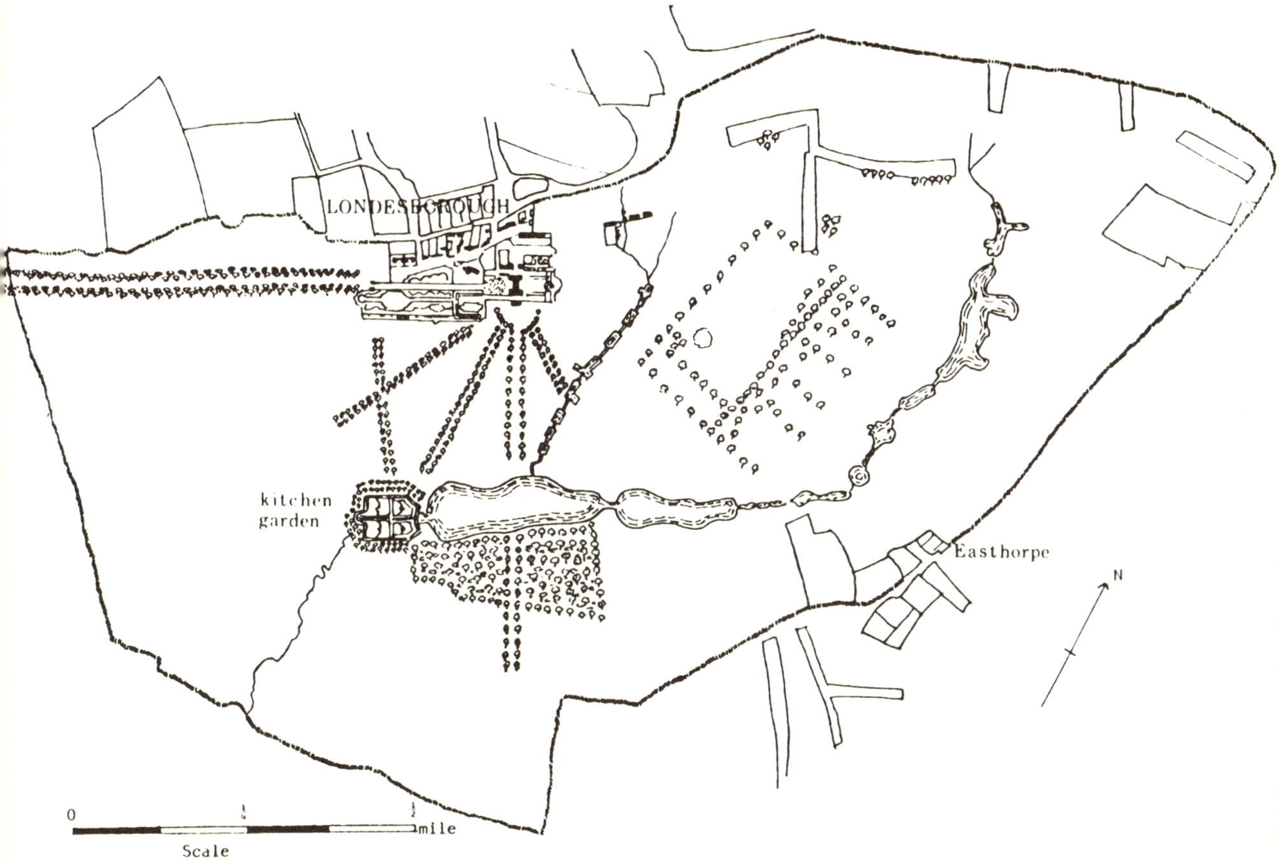

Londesborough Park, 1739, redrawn from an estate plan by Thomas Pattison, reproduced from Neave (D.), Londesborough *(1977).*

This garden layout on the west side of Londesborough Hall (p. 76) bears a great similarity to that shown on a plan of 1737, made by the engraver and surveyor John Rocque, of Sir Charles Hotham's garden at South Dalton, near Beverley (p. 77). Sir Charles and the Earl of Burlington were friends and were also related through marriage, and it is possible that while visiting Londesborough Sir Charles watched the making of the garden and decided that he would like a similar one himself. Knowlton must have been much involved in the planning and making of the Londesborough garden, but he also assisted Sir Charles with his, for we read in a letter he wrote to Samuel Brewer, dated 24 August 1729, that he had been to Sir Charles Hotham, and spent 'most of y^e^ afternoon with him about a new designe for a garden & padock &c.'(1)

Thomas Pattison's estate plan of Londesborough Park of 1739 (above)

(1) *See* below, p. 103.

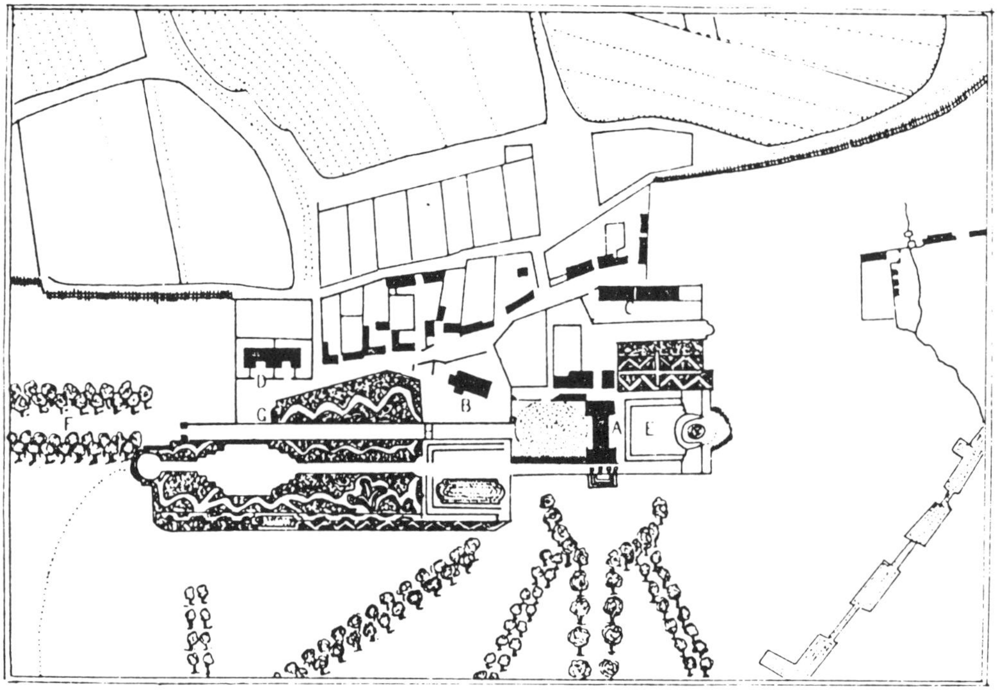

A The Hall
B The Church
C The Stables
D The Hospital
E The Fountain Garden
F The Long Avenue
G The Wilderness
H Garrick's Walk

Londesborough: plan of gardens, hall and village, 1739, redrawn from an estate plan by Thomas Pattison, reproduced from Neave (D.), Londesborough *(1977).*

shows two chains of irregular shaped ponds, or, as Knowlton called them 'canals'. The larger group cascaded down from the north-east corner of the park to the very large canal, known as the Great Pond. The other chain started much nearer to the house, it was shorter and the canals were smaller. As in the case of the larger chain it was situated to the east of the house and ran into the Great Pond. It is possible that Burlington was referring to the smaller chain when he informed Lady Burlington, in a letter dated 2 November 1739:(1)

'...as to my works in the park, this time of the year wou'd only admit of doing so much, as to drain that part of it which is next the house, this will be compleated to-morrow, and on tuesday I think to begin my journey. Burward has done so well, that when last week a man wou'd have sunk up to his middle, a horse may now go dry, he promises to be here in february, and to finish every thing himself before he leaves it, the inhabitants of the park, are already very happy in the large range they now enjoy...'

(1) Lord Burlington's correspondence, preserved at Chatsworth.

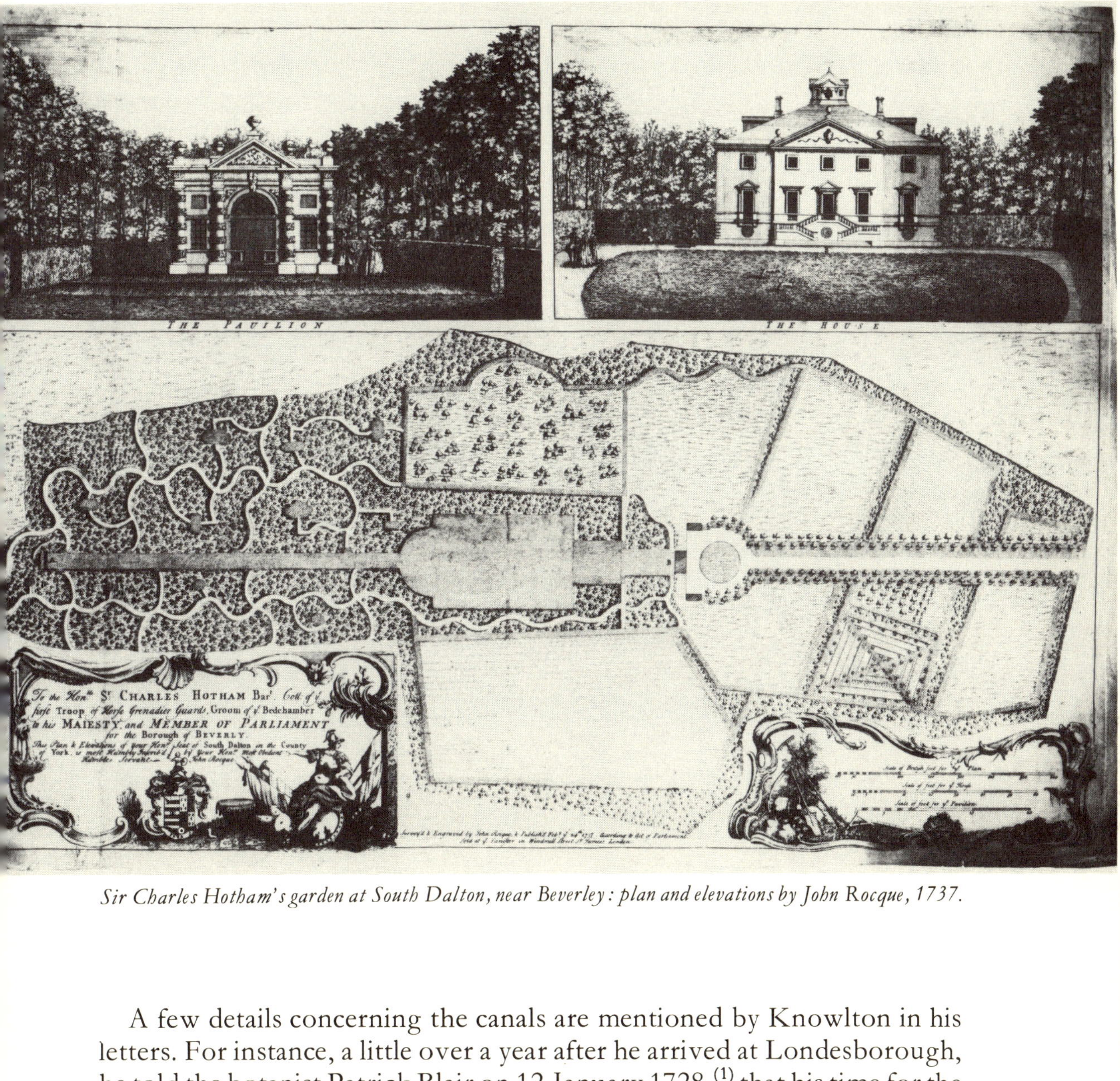

Sir Charles Hotham's garden at South Dalton, near Beverley: plan and elevations by John Rocque, 1737.

A few details concerning the canals are mentioned by Knowlton in his letters. For instance, a little over a year after he arrived at Londesborough, he told the botanist Patrick Blair on 12 January 1728,(1) that his time for the most part had been devoted 'to planting & making woods water &c ye beautifull ornament of a fine seat'. On 12 April of the same year he informed his friend Samuel Brewer(2) that fishing was a diversion he much loved, besides he had made 'severall peace of watter so fish[ed] to stock y^m

(1) *See* below, p. 85.
(2) *See* below, p. 94.

with.' Subsequently, he told the same botanist, on 8 February 1730[(1)], that he had just as many canals in the park as Dr Richard Richardson had at Bierley, and that they were all finished and full of water. However, he was then in the process of making two more and when completed he would have a fine chain of them.

It appears that Knowlton was here referring to the larger chain, the last two canals of which were much bigger than the rest. The end one, the Great Pond, was made to adjoin the kitchen garden. It is not known when these two stretches of water were completed but as we shall see later when the Great Pond was being excavated the Roman road that runs through the park was found. Knowlton informed the antiquarian Francis Drake of the discovery, and the latter included a description in his *Eboracum or, the history and antiquities of York*, published in 1736. It seems reasonable to assume that these two canals were those that Knowlton was in the process of making when he wrote to Brewer in 1730, and that the Great Pond was completed about 1732 or a little later.

The enthusiasm for sylviculture begun by the publication of John Evelyn's *Sylva* in 1664 gathered momentum during the eighteenth century, and many landlords planted not only to beautify their grounds but also to procure timber as a form of investment. Pattison's estate plan of Londesborough Park in 1739 (p. 75) shows that at that time there was a large plantation south of the Great Pond and that landscaping was also in progress on the east side of the park. Furthermore, apart from the semi-circle of trees from which a wide avenue led into the park, first featured in Kip's engraving, there were five further avenues bordered by trees. Two were added to the semi-circular group of trees, one going south-east to the small chain of canals and the other south-west to the kitchen garden. A second avenue to the kitchen garden started from the south-west end of the pleasure garden, and from the opposite end of that garden an avenue went into the park in a south-westerly direction. Finally, a fine wide avenue with double rows of trees went westwards from the Yew walk. As will be noted later, Burlington wished to continue this avenue to join the York road. To do this he had to obtain the permission of the owner of the adjoining estate, namely Sir Marmaduke Constable of Everingham, and this was granted in June 1742.[(2)]

In the north wall of the kitchen garden at Londesborough is a gateway built from a design, dated 1735, by the Earl of Burlington. It is of special

(1) *See* below, p. 120.
(2) *See* below, Appendix A, p. 289.

interest as being the only architectural feature known for certain to have been designed by the Earl for his Yorkshire country house. The design is now preserved in the Library of the Royal Institute of British Architects in London.

Unfortunately, Knowlton gives little information in his correspondence regarding the planning and planting of the park. His chief interest appears always to have been in the native flora and in the growing of exotic flowers and fruits. However, according to his own writings when he first went to Londesborough his time was mostly spent in tree planting, certainly up to 1739, when Pattison made his plan of the estate. Subsequently, although no plan or illustration appears to be extant showing what was accomplished between 1739 and Burlington's death in 1753, it is known that improvements in the park continued. For example, in a letter dated 8 March 1750 Knowlton wrote to Richard Richardson the younger that he had been much engaged during the winter in planting. On the 22nd of the same month he told Burlington's land agent, John Ferret, in a letter which will be transcribed later, of plans to make ridings, or walks, in the park.

The design of Londesborough Park and Gardens may be likened to that of many eighteenth century estates of the same period. The fashion which prevailed is well described by the novelist Tobias Smollett (1721–1771) when he wrote in his *Travels through France and Italy*:[1]

'In a fine extensive garden or park, an Englishman expects to see a number of groves and glades, intermixed with an agreeable negligence, which seems to be the effect of nature and accident. He looks for shady walks...for open lawns covered with verdure as smooth as velvet...for ponds, canals, basins, cascades, and running streams of water; for clumps of trees, woods, and wildernesses, cut into delightful alleys, perfumed with honey-suckle and sweet-briar, and resounding with the mingled melody of all the singing birds of heaven: he looks for plats of flowers in different parts to refresh the sense, and please the fancy...'

As stated earlier the kitchen garden at Londesborough was moved about 1730 from near the house to a site of four acres in the valley next to a water-mill. A surrounding wall was built in 1732. The Great Pond, probably completed about the same time, or a little later, is shown on

(1) Edition edited by Frank Felsenstein, 263 (1979) Oxford University Press. Letter dated Nice, March 5, 1765.

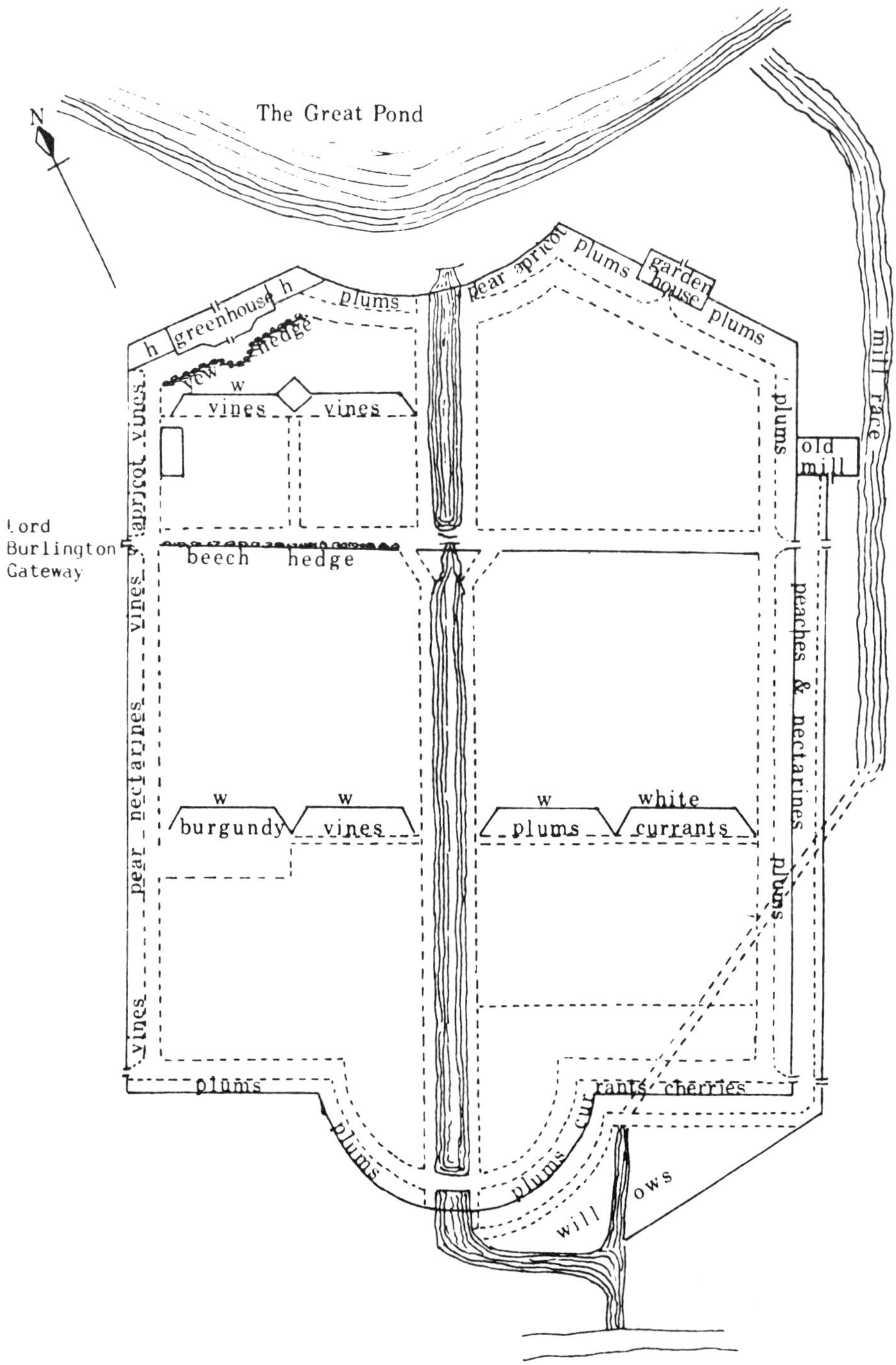

Kitchen garden at Londesborough, 1792, redrawn from a sketch plan by Thomas Knowlton the younger, reproduced from Neave (D.), Londesborough *(1977).*

Pattison's plan of 1739 at the east end of the kitchen garden (p. 75). The water-mill became the gardener's house and the mill-stream was diverted to run through the centre of the garden. The design of this garden, with its various structures and fruit trees, remained unchanged for many years after Knowlton's death. A note on a plan of the garden made in 1792 by the gardener's grandson, Thomas Knowlton the younger (p. 80), reads:

'All the walls round y[e] Garden in wretched need of pointing, & the upper part of the Brickwork perished, owing to their never having been finished with a coping, but only covered with thatch... all the Borders in a most impoverished State, and want renewing before young healthy Trees can be raised against the walls... they are at present furnished with... very small, puny, and sickly, Trees... The sorts are all well chosen by my Grandfather – of plumbs a very great variety.'

Besides Plums there were 'Burgundy' and 'Muscadine' vines, Pears, Nectarines, Apricots, Peaches, White Currants, and Cherries. The plan of the garden shows the position of several hotbeds and, situated in the north-east corner, facing south, was a greenhouse with a stove at each end, regarding which Thomas Knowlton the younger recorded:

'The Greenhouse & Stoves in very ruinous condition – have long been useless for y[e] purpose of cultivating Fruit, but their final destruction has long been laboriously protracted with my own hands, for the Conservation of a pretty & antient botanical Collection, by every expedient which could be thought of.'

When Knowlton first went to Londesborough he found no structures in which he could grow plants requiring heat. In fact it was several years before he had even a small stove. Philip Miller in his *Gardeners dictionary* of 1768 describes stoves as, 'contrivances for the preserving such tender exotic plants, which will not live in these northern countries whithout artificial warmth in winter', and in Knowlton's day there were in general use in England two kinds of such buildings: dry stoves and bark stoves.

A dry stove was heated by means of a fire from which heat passed into a flue. This wound from side to side at the back of the building until finally the smoke was discharged up a chimney, the plants were placed on shelves which were arranged in tiers, the rows rising above each other like the seats in a theatre, thus enabling all the plants to get an equal share of light and air.

These stoves afforded a dry heat and were principally used for the culture of tender succulent plants. A bark stove was a pit filled with tanner's bark and into this the more tender trees and herbaceous plants were plunged. The uniform, moderate, and moist heat, provided by the tan, enabled many species to be grown successfully which, before the use of tan, were considered impossible to cultivate in this country. Both dry and bark stoves were constructed with a high brick wall at the back, a lower one in front, and roofed with glass. Tanners bark was also used for hotbeds as distinct from stoves. These were paved to prevent the earth mixing with the tan and the sides bricked round. The whole was covered with moveable glass frames.

A greenhouse provided shelter for Orange trees and other plants that were too tender to remain out of doors during the cold months of the year. Such a building was always constructed to face due south in order for the plants to enjoy the maximum amount of sunshine. The back and two side walls were of brick while the front consisted mainly of windows to which were fitted shutters for use in cold weather. Some greenhouses had flues and a fire which could be lit in times of very severe frost, or when the weather was damp and foggy.

As already noted, both at James Sherard's garden at Eltham, and at Lord Burlington's at Londesborough, a stove was built at each end of the greenhouse so as to maintain plants of differing degrees of hardiness, regarding which Miller wrote in his *Gardeners dictionary*:

> 'I would... observe that the wing facing the south-east should always be preferred for the warmest stove, its situation being such, as that the sun, upon its first appearance in the morning, shines directly upon the glasses... warming the air of the house... after having been shut up during the long nights in the winter season... The entrance into the stove should be either from a green-house, the dry stove, or else through the shed where the fire is made, because in cold weather the frost-glasses must not be opened.'

According to the plan of 1792 there were two entrances to Knowlton's greenhouse, one at the front and another at the back, the latter leading perhaps to a shed for the furnace. When he designed his greenhouse and stoves Knowlton may have studied the pages of Miller's *Gardeners dictionary* where such structures are discussed and illustrated.

PART TWO

December 1726 to 28 November 1781

Gardener to the Earl of Burlington,
and the Dukes of Devonshire

The second part of this biography consists chiefly of quotations from Thomas Knowlton's letters written during the many years he lived in Yorkshire to correspondents including such eminent naturalists as Sir Hans Sloane, Dr Richard Richardson, Dr Patrick Blair, Samuel Brewer, Peter Collinson, John Ellis, and Emanuel Mendes da Costa.

The first known letter to be written by Knowlton, after he went to Londesborough as gardener to the Earl of Burlington, is to Patrick Blair (*c.* 1666–1728), a physician and Fellow of the Royal Society. He was the author of several treatises on medical and natural history subjects, including his *Botanick essays* 1720 [1719] and *Pharmaco-botanologia or, an alphabetical and classical dissertation on all the British indigenous and garden plants of the new London dispensatory* (1723–28). The letter contains a list of 31 English wild flowers, with details regarding when, and where, they were found.

TO DR PATRICK BLAIR

Londesburght Hall yorkeshire
about 15 miles from Hull Jan^{r} ye 12th 1727/8[(1)]
'Worthy S^{r}

it is now a Longe time since I have hard from you & Indeed from Mr Miller [Philip Miller, gardener at the Chelsea Botanic Garden] who ussed offten to Informe me of y^{r} wellfaire &c wch was a pleasure for me to heare, I remenber once you desired me to send you an accounts of such plants as I had found I mean those more uncom̄on's ones wch hear followes a short List of w^{t} I have obeserved

1. Anemonis geranifolio Ger[ard]: Em[aculatus]: 377 [i.e. Johnson's edition p. 377 (1633) of Gerard's *Herball*, sometimes referred to as 'Gerard emaculatus'] in plenty in a wood near Lutton Hooe in Bedfordshire 1714
   [*Anemone apennina* L., Blue Anemone; this species is considered to have been introduced]
2. Asarum of Dod[oens]: In a close not far from Sr Thos Seabrights of Beachwood [Flamstead, north-west of St. Albans] Hartfordshire 1716.
   [*Asarum europaeum* L., Asarabacca]
3. Astragulas incanus parvus purpureus nostras Syn[opsis] Icon 326 [i.e. Dillenius's edition pl. facing p. 326 (1724) of Ray's *Synopsis*] this grows in plenty one [on] most of woulds just above Estrope [i.e. Easthorpe] neare this place yorkeshire 1727
   [*Astragalus danicus* Retz., Purple Milk-vetch]

(1) Manuscript preserved in the Hunterian Library, in the University of Glasgow.

4. Bulbosa juncifolio growes plentifuley one [on] ye Rocky mountains in ye Island of Garnsey where observed (when Dr Dugulus [i.e. Dr James Douglas] sent me over about ye Garnsey Lilles [*Nerine sarniensis* (L.) Herbert]) & Brought many rootts & seeds wch M^r fairchilds [i.e. Thomas Fairchild, the Hoxton nurseryman] had parte & ye D^r a good quantity 1726
[*Romulea columnae* Sebast. & Mauri, Sand Crocus]
5. Chirstophoriana of Ger[ard]: this I found in Grate plenty riding from Slingby [i.e. Slingsby] one [on] ye Left hand & a rocky banck & about ye midway to Hoveingham yorkshire 1727
[*Actaea spicata* L., Baneberry, Herb Christopher]
6. Cotula foitidissima flore pleno found in ye 1711 neare baldoke in Hartfordshire & since neare Hull att a town cald Hasle [i.e. Hessle on river Humber about four and a half miles S.W. of Hull] 1727
[*Anthemis cotula* L., Stinking Mayweed]
7. Campanula perseij folio I found it in plenty united into ye pasture as one go's from pettworth to meadhust [i.e. Midhurst] in Sussex 1725 & afterward showed to Dr Langwith a curiouse Gentleman of that town who was Diffident & Brought of it Home & planted by ye other in his Garden & could obeserve no deferance I likewise brought some of it up to Mr Miller att Chelsey but I think it deyd—
[*Campanula persicifolia* L.; this species is considered to have been introduced]
8. Campanula maxima foliis lattissimus this growes plenty in Sr Charles Hothom's woods att south dorton [i.e. South Dalton] Yorkeshire [in East Riding]
[*Campanula latifolia* L., Large Campanula]
9. Campanula vid[e] Cantabury bell att Scorbrought [i.e. Scorborough] all about ruins Sr Charles Hothoms—
[*Campanula medium* L., Canterbury Bell.; this species is considered to have been introduced. At the time of Blair's letter Sir Charles Hotham lived at South Dalton, but until 1707 the Hotham's family seat had been Scorborough in the hamlet of that name, a few miles north of Beverley]
10. Digitallis flore albo this I see growing one [on] ye Island of Sark a League from Garnsey 1726
[A white form of *Digitalis purpurea* L., Foxglove]
11. dentaria major mathiolus in grate plent[y] neare fountians abby [Fountains Abbey] yorkshire 1727 & Likewise in att a wood near prison in Hartfordshire 1711 & in a wood att chiselhust in kent ye town I was Born att—
[*Lathraea squamaria* L., Toothwort]
12. gentiana palustris angustifolia C[aspar] B[auhin] s[ive] Calathiana violett this growes plentifully in our Sawampis com̄ons [swampy com-

mons] at Home [Londesborough] & Everingham neare me yorkeshire this plant is not about London nor in any of thire gardens
[*Gentiana pneumonanthe* L., Marsh Gentian]

13. Gentiana pratensis flore albo s[ive] Dwarfe autumnall Gentiana this I have found betwen Londsburght & Burham [Burnholm, now called Nunburnholme, north west of Londesborough(1)] grate plenty & gathred seeds 1727
[Possibly a white form of *Gentianella amarella* (L.) Börner, Felwort]
14. Lathrus sativus Like ye co͞mon in gardens only ye flower smaller & in plenty neare to battel bridge (one [on] bothe sides ye road one [on] ye other side ye heges more plenty) wch is between Aston and benington Hartfordshire 1725
[*Lathyrus latifolius* L., Everlasting Pea; this species is considered to have been introduced. There is no record of any road bridge having existed between Aston and Bennington before the present century, and the only footbridge shown in 1840 is nowhere given a name.](2)
15. Lentibilaria Riv[inus]: s[ive]. milefolium palustre galericum one [on] bothe sides ye road in Walling fen from Cave to Howden yorkshire 1727
[In Rivinus (A.Q.), *Ordo plantarum, quae sunt flore irregulari monopetalo* (1690), plate 79 is entitled 'Lentibularia', and in Johnson's edition (1633) of Gerarde's *Herball* is a woodcut on p. 828 entitled 'Millefolium palustre galericulatum'. Dr Peter Taylor of the Royal Botanic Gardens, Kew, has confirmed that both figures are of *Utricularia vulgaris* L.]
16. Lychmachea Chamaenerion flore purpureo in a wood belong[ing] [to] Sr Thos Seabright's beach wood [i.e. Sir Thomas Seabright's property called Beachwood in the parish of Flamstead, Hertfordshire]
[*Epilobium angustifolium* L., Rosebay Willow-herb]
17. Mespelus spinoso fructu alba white Hawfrute in Enfeild chase [Middlesex] ye sid[e] next to north Hall co͞mon & ye same time brought seeds cutings & 2 young plants & gave seeds to Mr Duboys [possibly Charles Dubois of the East India Company and the owner of a noted garden at Mitcham in Surrey] & to M^r^ fairechilds (Thomas Fairchild, nurseryman at Hoxton] youngs plants & cuting to forboys [possibly Robert Furber, nurseryman at Kensington]
[Possibly *Crataegus oxyacantha* L. var. *leucocarpa,* recorded by Loudon (J. C.) in his *Arboretum et fruticetum britannicum,* **2**, 832 (1838) as a variety of the Common Hawthorn with white haws, now a variety of *C. monogyna* Jacq.]

(1) In a map of East Riding in Camden's *Britannia,* **2** (1722), two places are shown within a short distance of each other, namely, Burnholme and Nun-Burnholme, now both called Nunburnholme. At Nun-Burnholme was a small Benedictine nunnery.
(2) Information supplied by the Archivist, Hertfordshire Record Office, in *epist.* 12.xi.80.

18. Mespilus Laurij folio & Hortenses this growes plentyfull in all y[e] heges rowes [hedgerows] & rocky mountains in Gurnsey 1726 y[e] frute is smaller y[n] ours in y[e] gardens
[*Mespilus germanica* L., Medlar, this species is considered to have been introduced]
19. Myrrhis alter parva. S[ive]. Sweett cissely att fountains abby 1727 yorkshire & markett cell [near Luton] Hartfordshire(1) 1714
['Myrrhis altera parva' is normally thought to be *Chaerophyllum aureum* L.; however, the English name and the locality strongly suggests that Knowlton was meaning *Myrrhis odorata* (L.) Scop., Sweet Cicely]
[Knowlton gives no entry 20 or 21]
22. Myosotus scorpioides latifolia hirsuta R Sy Icon & D[escription]: 228 [i.e. *See* Ray's *Synopsis* edited by Dillenius plate 9 description on page 229, not 228 (1724)] this growes plentyfuly neare fountains abby yorkshire & is y[e] pritys plant I ever see in my Life
[*Myosotis sylvatica* Hoffm., Wood Forget-me-not]
23. pshudo narssisus plenty 1000 of y[m] in daffadowndilly wood Sussex near pettworth
[*Narcissus pseudonarcissus* L., Wild Daffodil]
24. orchis plamata [i.e. palmata] s[ive]. froge orchis one [on] y[e] woulds above Killick House near pocklington [possibly 'Killick' should read Kilnwick, a property just over a mile north east of Pocklington]
[*Coeloglossum viride* (L.) Hartman, Frog Orchid]
24. [bis] periclyminum folis Qurcinus [i.e. quercinis] found in a hedge betwen Hitchen & Wimmondly [i.e. Wymondley] Hartfordshire 1711
[*Lonicera periclymenum* L., Honeysuckle]
25. plantago pannicula sparsis between Hitchen & preson Hartfordshire 1711 & now 1727 betwen north cave & Clife [i.e. Cliffe] yorkshire
[Both North Cave and Cliffe are in the East Riding, south of Market Weighton].
[Probably a monstrous form of the Greater Plantain, *Plantago major* L. forma *ramosa* Pilger]
26. Primula veris flore rubro one [on] wharfe side near badger forge by ottly [Otley on river Wharfe 6 miles east of Ilkley] yorkshire.
[*Primula farinosa* L., Bird's-eye Primrose]
27. Seedum mynus circinato folio s[ive] azoides dasyphyllon [i.e. *Aizoon dasyphyllum*] plentyfuley one [on] all y[e] walls & y[e] housetope of markett cell neare markett streett [In the vicinity of Merk-gate or Market Street, about three and a half miles south west by south from Luton, was Market Cell, a Benedictine nunnery founded in about 1145 (cell = a

(1) For details of Market Cell *See* below, Entry 27.

monastery or nunnery). In the time of Edward VI the monastic buildings were converted into a mansion which was called the Priory]
[*Sedum dasyphyllum* L., Thick-leaved Stonecrop]

28. Cotyledon vera tuberosa Radice J[ean] B[auhin] I never see this plant near London no otherways y^{n} In gardens & what was planted by Mr Dandridge not fare from Islandton [Islington?] nor any whare else but in ye Island of Gurnsey 1726 one [on] ye walls as Inclose ye feilds 1000 of 1000s I may say 100000—
[*Umbilicus rupestris* (Salisb.) Dandy, Pennywort, Navelwort]
29. Nidus avis s[ive] birds nest plenty ofly parke [Offley Park, Sir Henry Penrice's estate, near Hitchin] Hartfordshire 1714
[*Neottia nidus-avis* (L.) Rich., Bird's-nest Orchid]
30. Epimedum s[ive] baren wort is found in Grate plent[y] one [on] ye sid of steep rock neare Bingly [about five and a half miles north west of Bradford, on river Aire] by a garden^{r} of y^{t} town as ye worthy Dr Richardson Informes who upon notice therof sent his sarv^{t} with ye garden^{r} to ye very spott where was not less y^{n} a enough it to Load a horse. [*Epimedium alpinum* L., Barren-wort; this species is considered to have been introduced]
31. Likekewise [i.e. likewise] ye Campanula pantagona [i.e. pentagonia] of mor[ison]: without ye town plentifully one [on] ye mud walls &c. [*Legousia pentagonia* (L.) Druce; this native of the Aegean region and eastern part of the Balkan peninsula appears to have been cultivated in English gardens at this period]

thes S^{r} are ye most metearell [material] ones y^{t} I have observed to grow wild – I should be glad to know if another edition of ye Botanick Essays is yett published [the work was first published in 1719(1)] & what no^{rs} [numbers] of y^{r} pharmaco-Botanologia you have published [this appeared in seven parts or decades during 1723 to 1728 inclusive] for you must know I Live now a grate distance from any twon [town] & very seldom see any new's papers tho now & y^{n} mett a freind as ye worthy Dr Richardson & Sr Henry Goodrick(2) who are grate Lovers of Botany & gardining besides thers a physion att Hull y^{t} stuided physick att Leyden who I was when he was there very much aqueinted with he was then a grate Lover & observer of plants & nature I was but about a monthe ago Informed of his being thire [i.e. at Hull] his name is Dr Chambers who I in a short time hope to see & hope to have ye pleasuer of seeing you some time next springe or

(1) The title page bears the date 1720, the work, however, was apparently published in 1719 *See* Henrey (B.), *British botanical and horticultural literature before 1800*, etc., **2**, 49 (1975).

(2) For details regarding Sir Henry Goodricke *See* below p. 97.

sumer[1] for I exspect to be in Lincolonshire att Esq^{rs} Gores[2] so will if possable make you a visitt, but in ye meantime if y^{r} affairs should call you this way I mean towards york wch Londesburght would be in your way I should be Extreemly glad to see you & should take it as a faver. Mr Miller was this Last somers in Holand as M^{r} fairchilds [the Hoxton nurseryman] sent me word & he brought a grat many plants from thence; I hope y^{r} garden stoves incress to a Large collection wch in this part would be an agreeable sight to a curious person I have a few seeds sent me this winter & Likewise Brought some few out of Holand 1726 when I was thir for M^{r} fairchilds I Likewise have a fine Large stript aloe mucronato americana [probably *Agave americana* L. cv. Variegata, Century Plant] & 3 or 4 of ye Echinomellocactus minor [probably *Mammillaria simplex* Haw.] of ye paradisus Batavus [by Paulus Hermannus, p. 136 (1698)] & some others but thos are ye most s[c]arce... ye worthy Dr Richardson was so good to offer of his collection of all ye sorts he had Dupicates of if his Ldshipe would buld a stove whech he tould me he would when he sent me down [to Londesborough?] & ever since I have ben in this cuntry my time for ye most part has ben devoted to planting & making woods water &c ye beautifull ornament of a fine seat so have run thro y^{t} shall have more Leasure to Look after ye botanick afairs wch I hope in time to have a prity colection wch S^{r} bring me to a conclusion & to tell you I am with Grate Respects y^{r} most humble & obydant Sarv^{t}
Thos Knowlton
I hope this will find you in good health & that youl be pleased to faver me with a Line to heare of wellfair wch will be a perticular pleasure for me to heare of you may Direct to me att the Rt Honble ye Earl of Burlingtons att Londesburght neare markett weighton yorkeshire by ye way of york
NB if you want a Little mimosa seed of a very good kind I am ready to suply you if you can inform me ye way to send it – I should be glad to be informed who was ye translater & editer of Mr Turneforte Botanical Instutions I mean y^{t} one vol y^{t} is published it was a pity it was not finished I have heard said or Least Imagined by some y^{t} it was of y^{r} performance...'

The above passage refers to *The compleat herbal: or, the botanical institutions of Mr. Tournefort*, published in London in two volumes (1719–30). It

(1) Blair moved from London in 1720 to Boston in Lincolnshire, where he remained until he died in February 1727/28, only a short time after Knowlton wrote to him.

(2) It has not been possible to find any information concerning 'Esquire Gore', but he may have been related to Elizabeth, daughter of Sir Humphrey Gore, of Gilston, nine miles north east of Hertford, Hertfordshire, who in May 1714 married Sir Henry Penrice. Sir Henry through his marriage acquired Offley Place, the property in Hertfordshire, where Knowlton worked for a time as a gardener *See* above p. 35 and *The Victoria history of the counties of England. Hertfordshire,* **3**, 40 (1912).

originally appeared in parts, the first in 1716. The work is a translation from the Latin edition (1700) of Joseph Pitton de Tournefort's *Elémens de botanique*, with 'large additions from Ray, Gerarde, Parkinson, and others'. This translation is sometimes ascribed to the botanist John Martyn (1699–1768), but there is no supporting evidence for this attribution.(1) Martyn became acquainted with Blair as a young man and he is said to have looked upon him 'as the most intimate friend of his early life'. After reading Knowlton's letter we may ask ourselves whether Blair was the translator of the early numbers of *The compleat herbal*, and whether after his death in February 1727/28 the work was perhaps completed by Martyn.

Knowlton closed his letter as follows:

'... ye chamaepericlymenum [*Chamaepericlymenum suecicum* (L.) Asch. & Graebner, now called *Cornus suecica* L., Dwarf Cornel] as M^r [Robert] wood [Curator] of Edenburgh [Physic garden] some time since Informed Dr Richardson of ye grat plenty about 6 or 7 miles from y^t city(2) [and] one [on] chevett-hill [Cheviot Hills, range in Northumberland forming boundary between Scotland and England] some 100's of this plant grew he promised but non[e] has ben yett Recd I should be glad to know if you have ever seen ye plant in that garden [at Edinburgh] I Remember I once see it at Leyden Physick garden in flower & it was much Like to asassimum flower [*Sesamum orientale* L.,(3) Sesame].'

Most of the Knowlton correspondence extant at the present day is preserved in the Library of the Department of Botany in the British Museum (Natural History) and in the Bodleian Library, Oxford. In the former library are letters addressed to the botanist Samuel Brewer and in the latter are those to Dr Richard Richardson and to Richard Richardson the younger. The first known letter written by Knowlton to Brewer is dated 1728, and after this he continued to correspond with his friend until shortly before the latter's death.

Of the early history of Samuel Brewer (1670–1743), Pulteney recorded:(4)

(1) *See* Henrey (B.), *British botanical and horticultural literature before 1800*, etc., **2**, 50–51 (1975).

(2) According to Mr Wood *Chamaepericlymenum* grew near Edinburgh in the Pentland Hills. *See* below p. 112.

(3) *See* Saldanha (C. T.) and Nicolson (D. H.), *Flora of Hassan district Karnataka, India*, 562 (1976).

(4) Pulteney (R.), *Historical and biographical sketches of the progress of botany in England*, etc., **2**, 189 (1790).

'He was originally of Trowbridge, in Wilts, in which county he had a small estate. He was engaged at one time in the woollen manufactory of that place; but, I believe, proved unsuccessful in business.'

The reason of Brewer's failure in business may have been due to bad debts by a partner.[1] Brewer was a keen botanist and collector. In 1726 he spent two months on a botanical journey in the company of J. J. Dillenius. From Trowbridge, in Wiltshire, the two men went by way of Somerset, Gloucestershire, and Shropshire to Wales, including Anglesey. At the end of this journey Dillenius returned to London, while Brewer spent the rest of 1726, and the greater part of the following year, in North Wales, collecting around Bangor, Snowdon, and elsewhere. During this time Brewer sent Dillenius parcels of plants.[2] Dillenius thought very highly of Brewer as a botanist, and on 30 January 1727 he wrote to the famous Swedish naturalist, Carl Linnaeus:[3]

'I am sure I shall never meet with a better searcher, especially for mosses. When we travelled together in Wales, he would stop in the worst weather and most violent rain to pick up mosses.'

Shortly afterwards, the same authority told Linnaeus,[4] that Brewer had 'a particular genius for plants, and had he been bred a scholar, he would have done great things.'

In 1728, Brewer was residing in Yorkshire, where he remained until he died. Knowlton addressed a letter to him, dated 12 April 1728, 'To Sam.^{ll} Brewer Esqr Lodging in Bradford or else att ye worthy Dr R. Rich^{d}son's att north Bierly yorkshire at ferry bridge'. After this letters were addressed to 'his Lodging in Bradford yorkshire', and from 28 June 1733 until 15 December 1741 to 'North Bierly neare Bradford...'

At Bierley, Brewer collected plants, insects, fossils, and other subjects of natural history interest. He also cultivated a garden where he grew plants

(1) *See* below, p. 180.

(2) For details of the expedition taken by Dillenius and Brewer to North Wales, Brewer's journeys in North Wales during the subsequent twelve months, and the names of some of the many plants collected *See* Hyde (H. A.), Samuel Brewer's diary: a chapter in the history of botanical exploration in North Wales, *Bot. Soc. and Exchange Club of Brit. Isles. Report for 1930*, **9**, Part 3 (August 1931) Supplement, and Ellis (G.), Plant hunting in Wales, *Bull. National Mus. Wales,* 15–25 (Spring 1974).

(3) Smith (Sir J. E.), *A selection of the correspondence of Linnaeus,* etc., **2**, 144 (1821).

(4) Smith (Sir J. E.), *op. cit.* **2**, 146.

both native and exotic. Furthermore, in June 1733, Knowlton heard that his friend was compiling a list of his observations which he intended to publish under the title of 'A botanical guide',[1] but this work he never completed.

Brewer sent botanical specimens to various naturalists, including Sir Hans Sloane who told Dr Richard Richardson in a letter of 13 August 1734:[2]

'I desire you to give my service to Mr. Brewer... I would be glad if he sent me up specimens of the plants, insects, and fossils, or other natural productions of his neighbourhood. He hath leisure, is well skilled in drying, pasting, &c. and will receive from me some consideration for his trouble.'

During his life Brewer suffered various trials and tribulations including, as will be seen later, monetary losses, and worries with regard to the behaviour of his son and daughter[3]. He seems to have been often in need of money and dependent to a certain extent on the generosity of his fellow naturalists. Dr Richardson, considered by one writer to be always ready to assist his friends, was especially kind to him[4]. Brewer did not always enjoy good health and towards the end of his life was a very sick man. In an undated letter to Sir Hans Sloane which bore the post office date stamp '29 Feb', the year being almost certainly 1740, he wrote:[5]

'I... cannot expect long to be ffree or even to be able to do anything more towards compleating my collection or to publish the observations Ive made as intended, [the 'observations' were to be published as the 'Botanical guide'] so... I have nothing so much at heart as to the disposing of it all to gether as it is, wch consist of Plants Foserls eggs & insects, & could heartedly wish you could find... a customer for it.'

The date of Brewer's death is always given as 1743 but on what authority the present writer has been unable to discover. It is recorded by Dorothy, wife of Richard Richardson the younger,[6] that he 'was buried in

(1) *See* below, p. 132.
(2) *See* Nichols (J.), *Illustrations of the literary history of the eighteenth century,* etc., **1**, 288 (1817).
(3) *See* below, pp. 180–81.
(4) Nichols (J.), *op. cit.* **1**, 415 (1817).
(5) Sl. MS. 4076 fol. 341 in British Library.
(6) Nichols (J.), *op. cit.,* **1**, 288 footnote (1817).

the yard close to the East Wall at Cleck-Heaton Chapel, near Bradford'. According to the Reverend R. M. Wilson, Vicar of St Luke's and Vicar of Whitechapel, Cleckheaton(1) 'Of the grave yards in Cleckheaton only the one at Whitechapel is so old as to have been in use in the 1700's'. On examining the graves close to the east wall of Whitechapel parish church the Vicar found no grave in the vicinity which might conceivably be Brewer's though there are several graves in that general area which are of great age but whose inscriptions are illegible. On 8 June 1743, Knowlton sent a box of plants to Richard Richardson the younger, together with a letter and a plant he asked him to give to Samuel Brewer.(2) This plant Brewer had written for the previous month. He may have died soon after this was sent as subsequently there is no further correspondence between the two men.

Brewer lived in the same neighbourhood as the physician, antiquary and naturalist, Richard Richardson (1663–1741) F.R.S. who had an estate at North Bierley. Frequent mention of him is found in Knowlton's letters where he is often referred to as 'the Dr.'. Richardson had an extensive medical practice. He corresponded with Sir Hans Sloane, President of the Royal Society from 1727 to 1740, and with many outstanding botanists and naturalists of the day, as well as with persons noted for their antiquarian and classical interests. He formed a valuable library of botanical and historical works. His garden was outstanding for its large collection of plants, both indigenous and exotic, and he was among the first in the north of England to build a hot-house.(3) Furthermore, he was a keen botanist and travelled in England, Scotland and Wales in search of rare botanical specimens.

We must now return to Knowlton's correspondence beginning with a letter he wrote to Samuel Brewer.

TO SAMUEL BREWER

Londesburght April ye 12th 1728(4)
'Kind S^r^
Y^rs^ I Recd & am very glad to heare of y^r^ wellfair & Likewise of y^r^ being

(1) In *epist.* 11.xi.80 and 2.xii.80.
(2) *See* below, p. 202.
(3) *Extracts from the literary and scientific correspondence of Richard Richardson,* etc. [Edited by Dawson Turner], XXX (1935) Yarmouth.
(4) Knowlton correspondence preserved in the Department of Botany of the British Museum (Natural History).

Dr Richard Richardson (1663–1741) of North Bierley, a keen botanist and the owner of an outstanding garden. He corresponded extensively with Sir Hans Sloane and many other naturalists and antiquarians.

in this country [i.e. Yorkshire], I should be very glad to see you att this place for 5 or 6 days if it will sute with y[r] convenance so y[e] sooner y[e] better haveing now opertunity to Recd a fraind before his Lordship visitts this place it is to soon to mak any outs in search of plants by reson y[e] Spring is so Latte but as soon as it is (& my place will give Leave) I intend God willing to vissitt Craven whare I have an Invitation &c it is by repourt a fine place for plants.'

Included among the Earl of Burlington's vast estates was Bolton Abbey in the deanery of Craven,[1] in the West Riding, and it may have been there that Knowlton had been invited.

Knowlton continued by writing of 'ye astragalus Incana &c.', this being an abbreviation for the plant referred to by Ray[2] as 'Astragalus incanus parvus purpureus nostras', and now named *Astragalus danicus* Retz., Purple Milk-vetch. It will be remembered that Knowlton told Blair[3] that this plant grew 'in plenty one most of woulds just above Estrope [Easthorpe]'. In writing to Brewer about the same plant Knowlton told his friend:

'... ye astragalus Incana &c we have grate plenty... but now not ye Least apearance of, tho I can very Easly suply'd ye good D^r a Likewise what others I have y^t he has Lost... would be att his or y^r sarvice being never selfish y^t way.'

Knowlton was afraid that at Londesborough he would not be able to grow many flowers, regarding which he wrote:

'... ye percariousness of sarvice & a multiplicity of other Bussness joyned with ye convesation of thos who are Lovers of planting & ye Desires of y^m to bend my Inclynation y^t way with thire good Incouragement has in a grate measure stivefled thos Botanicall thought y^t once was aspiring... but amidst all give me Leave to tell you I never Lived so happy as now in all ye course of my Life... thanks to the best of L^{ds} & frainds.'

The gardener informed Brewer:

'I this day am att york & from thence to ye worthy Sr H. Goodricks where I am to meett one of my good fraindss a grate Lover of planting Esq Darly &c.'

(1) *See* Whitaker (T. D.), *The history and antiquities of the deanery of Craven in the county of York... Third edition... edited by A. W. Morant,* etc., 497 (1878).

(2) *Synopsis methodica stirpium britannicarum... Editio tertia,* etc. [Edited by J. J. Dillenius], 326 (1724).

(3) *See* above p. 85, Letter to Blair (1728) Entry 3.

This was Richard Darly of Bishop Wilton, four miles north of Pocklington, in the East Riding. It is of interest to note that it was in the garden of Sir Henry Goodricke, Bart. (1677–1738) of Ribston, whose property was on the river Nidd, three and a half miles south east of Knaresborough in the West Riding, that the original Apple tree of the Ribston Pippen appeared. It is thought to have come from some pips brought there from Rouen in Normandy and planted in 1688(1). More is learnt concerning Goodricke's activities as a horticulturist in a letter of 9 February 1712/13, which he addressed to the naturalist James Petiver(2):

'It is to desire of you that if amongst y[r] rarities, you have any number of seeds fruits, or kernells of foreign and rare trees, especialy those that are hardy, I shall verily thankfully pay for em; my pleasure being to raise such things in hotbeds and preserve em with care; and I wou'd not rob you of any but what you have so many as you may readily spare a part, to one who will as readily supply you again, when any accident happens to yours, which I believe y[rs] are more subject to near London than we are here, where I my self take the cheif care of my curious trees. I have not yett been able to procure a tree of the true lotus [possibly *Celtis australis* L., commonly called the European Nettle tree](3) nor the Larch tree [*Larix decidua* Miller] both which Mr Evelyn says grow well in our climate, & may be raised of seeds(4); those seeds or any other exotics I doubt not to raise, I mean trees, for smaller plants are too numerous for me to attend; if you cou'd procure me a small tree, of each of those kinds I cou'd repay you with thanks.'

To return to Knowlton's letter to Brewer. In the following passage he may have been referring to the list of plants he included in his letter of 12 January 1727/28, addressed to Dr Patrick Blair, although by this time Blair was dead:

'In y[e] cat: to y[e] D[r] I forgott severall others as y[e] orchas bombely or

(1) Taylor (H. V.), *The apples of England,* 40 (1936).
(2) Manuscript preserved in the British Library Sl. MS 4065 f. 96.
(3) *Celtis australis* was cultivated by the herbalist John Gerard in his garden in 1596, under the name of 'Lotus arbor. Nettle tree', *See* Gerard's *Catalogus* of 1596 and 1599. The fruit of this species is sweet, and is said to have been the lotus of the ancients, the food of the Lotophagi (i.e. lotus eaters), a fruit so delicious that it made those who ate it forget their own country, *See* Bean (W. J.), *Trees and Shrubs hardy in the British Isles,* **1**, 567 (1970) and Loudon (J. C.), *Arboretum et fruticetum,* **3**, 1414 (1838). It has been considered by some that Goodricke may have been referring to *Diospyros lotus* L.
(4) Sir Henry Goodricke was here referring to John Evelyn's *Sylva* (1664), etc.

bumble orchas [*Ophrys apifera* Hudson, Bee Orchid] & another strange globelar ball of moss found about 3 fett deep in watter in grate quantitys it apeard att botum Like peples stons ye size of a tennes ball suprisingly odd to me; haveing never seen nor heard any thing of it befor, nor in the least taken notice of by ye people I shall be that way fishing in a short time so should be glad of y^r^ company... it being a devirtion I much Love besides have maid severall peace of watter so fish to stock y^m^ with haveing gott & bought all sorts but carpe, wch are not to be had in this country [Yorkshire].'

Later, we shall refer again to Knowlton's 'strange globelar ball of moss found... in watter', as, in 1729, he wrote both to Samuel Brewer and to Sir Hans Sloane about this green alga, now called *Cladophora aegagropila*.

At the time of writing his letter Knowlton intended to visit Fountains Abbey, near Ripon, in the West Riding where, he told Brewer:

'Lives a good senyor Romanet gentleman who has much presed me & any fraind I should bring to a night or too but indeed I very rare make... stays in any place but theres naturs temptations ye gentleman showd me in Aprill ye birds eye dentary major mat! [i.e. 'Dentaria major' of Matthioli,(1) now called *Lathraea squamaria* L., Toothwort] Lunaria minor [*Botrychium lunaria* (L.) Sw., Moonwort], Herba paris [*Paris quadrifolia* L., Herb Paris] Cardamina impatiance [*Cardamine impatiens* L., Narrow-leaved Bittercress] all wild just by ye celebrated Aby with divers others to tegious now to be numbed.'

The gentleman who invited Knowlton to stay a night or two may have been 'John Michel Messenger, esq., a worthy Catholic and antiquary', who lived in a venerable house adjoining the Abbey Mill, built out of the materials of the abbey.(2) Knowlton continued by informing his friend:

'I had a Letter ye other day from Dr Chambers of Hull who Lett me know ye Ill state of health of consall Sherrard wch I was very sorray for but death you know waits one all.'

This was William Sherard and Knowlton was correct in thinking that he would not live much longer – he died the following August. 'Dr Chambers

(1) The name under which the plant appears in Gerard (J.) *The herball... enlarged and amended by Thomas Johnson*, 1585 (1633).

(2) *See* Richard Gough's edition (1607) of William Camden's *Britannia*, etc., **3**, 57.

of Hull' was William Chambers (d. 1785), a native of that City and a particular friend of William Sherard and his brother James who considered him 'very curious in botany and all other parts of physick'.(1) Knowlton himself, it will be remembered, referred to him as 'a grate Lover & observer of plants & nature'.(2) Chambers graduated on 11 August 1724 as Doctor of Medicine at the University of Leyden and in October of the same year was admitted an Extra-Licentiate of the College of Physicians of London. Subsequently he settled at Hull.

Knowlton closed his letter by once again pressing his friend to visit him at Londesborough:

'... give me leave once more to tell you if youl except such entertainment as my table afords you shall be extreemly welcome.'

In a postscript he added:

'Pray be pleased to give my most Harty Humble Sarvice to ye good D^r & his Lady & please to Lett him know by his ordur I Recd ye cutings of Ger. Sedums(3) &c for wch I aceknowledge myself much obidged & wish It was in my powere to retaleate Indeed when came from Holland & Gurnsey I had severall fine things & Large no^rs of my own I mostly gave to one fraind & other I corerespond with very few... now... except... my old fraind M^r fairchild [i.e. the Hoxton nurseryman, Thomas Fairchild] who... is most generous In his offers y^t way.'

Finally Knowlton informed Brewer that he would shortly be going:

'... towards the Honble ye Earle of Carlils & where ye Christophoriai [*Actaea spicata* L., Baneberry, Herb Christopher] grows &c.'

The Earl of Carlisle's property, Castle Howard, was in North Riding, a few miles west of Malton. Earlier, Knowlton recorded finding Herb Christopher in 1728, when riding from Slingsby to Hoveringham(4), Slingsby being a few miles north west of Castle Howard.

(1) Nichols (J.), *Illustrations of the literary history of the eighteenth century*, etc., 387 (1817), *See* letter, dated August 1723, addressed by William Sherard to Richard Richardson.

(2) *See* above p. 89, Knowlton's letter to Blair (1728).

(3) Presumably a reference to the 'sedums' included by Gerard in his *Herball* (1597) and in Johnson's editions of the work (1633, etc.).

(4) *See* above p. 86.

Twice in this letter of 1728 Knowlton told his friend that he would be very welcome at Londesborough. Brewer accepted the invitation and when writing, on 10 January 1730, to Sir Hans Sloane recounted:

'The sumar before the last I was wth m^{r} Knowlton at Londsburght who carried me 3 miles to shew me my Scarlet Alder [probably *Alnus glutinosa* (L.) Gaertner cv. Rubrinervia] y^{t} I first fetched in Wiltch [Wiltshire]'(1)

The finding of this Alder was mentioned in Dillenius's edition (1724) of Ray's *Synopsis*, as follows:

'Alnus vulgaris, sub-conis ligulis membranaceis rubris donata. The scarlet Alder. Found by Mr. Brower in the great Meadow below the cowhouse near Longleet...'

Further details concerning the 'Scarlet Alder' found by Brewer in Wiltshire are recorded below, where notice is also taken of a similar tree found in Yorkshire by Knowlton.(2)

TO SAMUEL BREWER

Londesburght oct^{r} ye 23^{d} 1728(3)
'S^{r}

Intend sunday come seven night to be att Bradfored where I hope to see you I have had thoughts a Long time but allways was Disapointed I Likewise shall proseed to marton wood if wether gives Leave in order to find Mr Listers museroms which...in ye transaction he extols so much tiss at ye Desire of Mr Stephen Switzer I persume into thos parts.'

Marton is in Craven in the West Riding of Yorkshire and Knowlton went to the wood at the request of the nurseryman and seedsman Stephen Switzer (1683?–1745). The work referred to was first read before the Fellows of the Royal Society on 20 November 1672.(4) Afterwards it was published in the Society's *Transactions*, Vol. **7**, No. 80, 5116–18, for 16 December 1672, under the title of 'A description of an odd kind of

(1) Sl. MS. 4051 fol. 166.
(2) *See* below, pp. 108, 155.
(3) Knowlton corres. British Museum (Natural History).
(4) Information received from Mr L. Townsend, Archivist at the Royal Society.

mushroom, yielding a milky juice, much hotter upon the tongue than pepper, etc'. The author was the eminent physician and zoologist Martin Lister (1638?–1712) who had seen these 'mushrooms' on 18 August 1672 when passing 'through Marton woods, under Pinno-moor in Craven'. He wrote of them, as follows:

'They were of a large size, something bigger than the ordinary red-gilled eatable mushroom or champignon, and very much of their shape, that is, with a perfectly round cap, or stool (as we vulgarly call it,) thick in flesh and with open gills underneath; having a thick, fleshy, not-hollow, and round foot-stalk, about 6 fingers breadth high above ground, and ordinarily as thick as my thumb. If you cut any part of this mushroom, it will bleed exceeding freely a milk-white juice.'

The fungus seen by Martin Lister was probably *Lactarius piperatus* (Scop. ex Fries) Gray, or *L. vellerius* (Fries) Fries.[1]

Knowlton continued his letter as follows:

'I hope ye worthy Dr Rich^{d}son's & Stanopes well... I had grate hopes I should have seen you in thes parts being so Long in this country [county] I supos you seen in ye newspapers ye deth of ye consall [i.e. William Sherard], in a few years it will be ours so shure it is.'

In a postscript he added:

'...his L^{d}Ship has been down twice this sumer & has ye Last time maid 5 weekstay... he very much pleas'd with all aditions maid wch is a satifacton I can please...'

TO DR RICHARD RICHARDSON

Londesburght Aprill ye 22^{d} 1729[2]

'Worthy S^{r}

According to my promis I have sent the following things which I hope will come safe to y^{r} hands & are as viz:'

(1) Information supplied by Dr Derek Reid, mycologist, at the Royal Botanic Gardens, Kew. It is interesting to note that in a section entitled 'Rare plants found in Yorkshire' in William Camden's *Britannia... Enlarged by...* R. *Gough*, etc., **3**, 98 (1789) there is an entry reading: 'Agaricus piperatus. Pepper agaric; in Marton wood near Craven'.

(2) Richardson correspondence preserved in the Bodleian Library, Oxford, MS. Rad. Trust. C.7.

There follows a list of twelve items which include:

'Stramonium flore duplici [possibly a double form of *Datura metel* L., Horn of Plenty]...a fine stript alloe [probably *Aloe variegata* L., Partridge-breast Aloe]...plantago panicula sparsa [*Plantago major* L., Great Plantain]...the Little Hedge hog [possibly *Aloe humilis* (L.) Miller, commonly called Dwarf Hedgehog Aloe or merely Hedgehog Aloe, native of Eastern Cape; alternatively possibly *Medicago intertexta* (L.) Miller, Hedgehog Medick, native of the Mediterranean region and Portugal, of which Philip Miller wrote: 'Snail Trefoil with a large fruit, whose spines point upward and downward, commonly called Hedgehog']...Comōne Gaile [*Myrica gale* L., Bog Myrtle, Sweet Gale]... & being I believe what will be exceptable all, wch I promis myself not to have seen in y^r^ garden, so I hope the more welcome I should be glad if it was in my power to Retaliate the singular favers Recd from you...NB...I had a letter from Buckingham House in St James parke & my fraind tells me ye double nasturtum [*Tropaeolum majus* L., Double Nasturtium] is Killd every-where about the town wch I am sorrey to heare but hope it will be preserved some.'

It will be remembered that Knowlton before he went to Yorkshire may have worked for a time in the garden at Buckingham House.(1) He often wrote of the Double Nasturtium and of his various efforts to cultivate the plant. Miller in his *Gardeners dictionary* (1768) states that the variety of *Tropaeolum majus* with double flowers is propagated by cuttings and that the plants must be given shelter in winter. It appears from eighteenth-century catalogues that it was always treated as a greenhouse plant.

At the end of his letter Knowlton said that he had intended to send the promised goods to Richardson by the nail man on the day he dated his letter, 22 April. However, the man did not arrive until 29 April and in the interval he added '2 sorts of very fine polyanthus & double primroses &c.'

TO SAMUEL BREWER

Londesburght July ye 3^d^ 1729(2)
'S^r^

I Recd y^rs^ & should have answer'd sooner but had thoughts of seeing you but now Haymaking is come & besids exspect to see his L^d^Ship every day so cannot come but promis myself ye pleasure about St Michellmuse [29

(1) *See* above, p. 65.
(2) Knowlton corres. British Museum (Natural History).

September]; y^{r} astragalus seeds will take care [to gather] haveing grate plenty neare as to other plants of the northe I have not any or att Lest very few...I Return thanks for all favers past & should be glad if I could oblidge & Likewise to see you I am sorry that Mr Sherrd [i.e. James Sherard] should medly with me when he knows can do no harm & Likewise how Little I vally it I am shure he had not ye Lest Reson if truth & honist meanings could be seen into but enought of that & I know very well wch is ye hapist man tho a poor gardener its plain he never was so by his temper nor I never a better plaise haveing not one thing I could wish to be mended or alter'd & have ye pleasure to pleas one of the Best noble man in the Kingdom who knows how to vally a good servant that maks me In every Respect so Happy; Indeed as to plants I have nothing more then pure nature to regail me that way but am as well content as ever my though being chefly taken up in ye Improvements of ye country as gardening Husbandry &c – wch is all att this time from S^{r} y^{r} most affection^{t} fraind & Humble sarv^{t}
Tho Knowlton
NB...I hop Dr Stanope is well pray give my Humble sarvice to him.'

Knowlton, it will be remembered, is said by William Sherard to have left his brother's employ as gardener at Eltham, in Kent, 'in a huff'.[1] The reason for this state of affairs, if indeed the statement was true, is not known. James Sherard, a keen botanist and collector of plants does not always appear to have been a very amiable man. The fact that he was unable to replace Knowlton with an equally skilled person may have done little to improve his temper, and it appears from the above letter to Brewer that he was now trying to make trouble for Knowlton at Londesborough.

TO SAMUEL BREWER

Londesburght August ye 24 1729[2]
'S^{r}
I Recd y^{r} very oblidging Letter for wch I Return my Kind thanks & am glad you was faver'd with ye sight of ye garden agreeable so much to y^{r} own desire the day I parted with you I waited one ye Hon^{bl} Sr Charles Hothom & spent most of ye afternoon with him about a new Designe for a garden & padock &c.'

(1) *See above*, p. 42.
(2) Knowlton corres. British Museum (Natural History).

It seems likely that the new design for a garden and paddock was for South Dalton, the property Sir Charles Hotham (1693–1738) owned, about five miles to the north west of Beverley, in the East Riding. Until 1707 the Hotham's family seat had been Scorborough, in the hamlet of that name about five miles north of Beverley, but in that year, during the time of the fourth baronet Sir Charles's father, the house was burnt down and not rebuilt, leaving the son to inherit in 1703 only a park at Scorborough.(1) Besides his country property Sir Charles Hotham also owned 'a large and spacious house' in the centre of Beverley.

Knowlton continued by giving a detailed account of what he did subsequent to leaving Sir Charles Hotham:

'... the next day went from hence to Hornsey where I walk one [on] the sea shore four miles in hops to find some old aqu^{ts} [acquaintances] but to my disopintment mett with nothing only ye Kaly spinosa [*Salsola kali* L., Saltwort] Anthyllis maritima lentifolia ['Anthyllis maritima,' *Frankenia pulverulenta* L., Powdery Sea-heath, is a species not now considered to occur in Britain where only *Frankenia laevis* L., Sea-heath is found. It is likely that Knowlton was confused about the name and that the plant he saw was 'Anthyllis lentifolia', *Honkenya peploides* (L.) Ehrh., Sea Sandwort, common all round the British Isles on sandy sea coasts] Eruca marina [*Cakile maritima* Scop., Sea Rocket] was all I found but had ye pleaser of Bathing for an hower after y^{t} dinde at my Inn & drank a Bole of punch & att 4 a clock seat out for Londsburght but from that time Raind all ye way home though before a very fine day. This day sevennight(2) I seat out for Whitby Lythe &c being about 40 miles from this place where I had the pleasure of seeing the allum works... butt could find no scarce plant... only ye Erica Baccifera [*Empetrum nigrum* L., Crowberry] but Riding from Lythe [a village three and a half miles north west of Whitby] one [on] ye sea shore had ye sigh[t] of such prodigious q.^{tity} of Eringo maritimy [*Eryngium maritimum* L., Sea Holly] & Rhamnus maritimus [*Hippophaë rhamnoides* L., Sea Buckthorn] y^{t} I was surprisd with ye grates pleasure & in my way home one [on] ye More's saw one of finest cascad[e]s I soer see before ye fall 30 yds deep & all naturall.
if I should wait one Sr Henery [Goodricke?] will think of you & belive you hardly saw one half of his trees they being scaterd up & down all round at distances(3)... I am very glad to heare you have so very good a London correspondence.'

(1) Stirling (A. M. W.), *The Hothams being the Chronicles of the Hothams of Scorborough and South Dalton,* etc., **1** (1918).
(2) A week, any period of seven consecutive days and nights.
(3) Regarding Sir Henry Goodricke *See* above, p. 97.

Attention has already been drawn to Knowlton's interest in books. In the present letter to Brewer he wrote:

> 'I wonder you have not gott M^r petivers works thirs a grate many good things Ley disposed up & down his voll: thre [that are] no whare to be had but of Sr Hans Slone who bought all his collections.'

The London apothecary and naturalist James Petiver (1664?–1718) had one of the largest and most varied collections of natural history specimens in England during the first quarter of the eighteenth century. These he used in preparing his many works, a number of which even during his lifetime became very scarce. Petiver's collections acquired by Sir Hans Sloane included his specimens and principal manuscripts, and in 1732 Sloane republished the 72 plates consisting of text and figures of plants comprising Petiver's *Catalogue of Mr Ray's English herbal.* Each figure has the name of the plant depicted together with references to Ray's *Historia* (1686–1704) and the *Synopsis* (1690). The work was of great value to botanists.[(1)]

Knowlton was always anxious to procure works containing plates which helped him to identify plants and other subjects. He liked to keep up to date with new publications, to know the price of a book, and whether it was well thought of. Moreover he was wont to praise or criticize books himself. After commenting on Petiver's works Knowlton asked Brewer: 'I should be glad to know if you have Barrelleo Icons by Jesew att paris 1714 folio & what prise & how you Like it.' The first of the two names mentioned refers to the botanist and Dominican Jacques Barrelier (1606–73) who was engaged upon a large botanical work at the time of his death. Subsequently, the copper-plates prepared for the project were collected together by Antoine de Jussieu (1686–1758), Professor of Botany at the Jardin du Roi in Paris, and published with accompanying descriptions in a work entitled: *Plantae per Galliam, Hispaniam et Italiam observatae, iconibus æneis exhibitae a* R. P. *Jacobo Barreliero . . . Opus posthumum, accurante Antonio de Jussieu*, etc. This work published in Paris in 1714 contains many illustrations, including over 1200 figures of flowering plants.

In this letter Knowlton again mentions the green alga *Cladophora aegagropila* that he wrote about to Brewer on 12 April 1728, as a 'strange

(1) *See* Henrey (B.), *British botanical and horticultural literature before 1800*, etc., **2**, 84–85 (1975).

globelar ball of moss'. He said:

'I... forgot to tell you how far I was oblidge to waide for the pillas (1) wch was allmost a 4th part of mile into ye meare or watter wch is not above 3 fett or 3 fett 8 Inches deepe.'

Later, the gardener wrote on this subject yet again. Meanwhile he continued his letter as follows:

'I thanke you for y^{r} generious offer of pertaking with you in thos plants Latly Recd its an obligation so grate will not be in my power to Retalliat such favers as I Know ye vally of y^{m} besides would be unkind of me in disiring without you had Duplycats & then a man is not allway safe in his Stock besid when Lost in this Remot Country will be very dificult to procure againe I bege you'l be so very Kind to give my most Harty Humble sarvice to ye worthy Dr Richardson & his Lady & hope they are well... I Recd his favers a Letter wch I ought to have acknowlegde sooner but had thoughts of seeing him before is ye reson of this omition if amoung y^{r} plants double nasturtum [*Tropaeolum majus* L.] be I should acknowlige it a very grate faver if I could gain a plant being first brought over by my self & the sooner sent so much ye better ye season declyning apace – if it is not in y^{r} hands, if eather Dr Rich^{d}son or Stanope, I wish you'd be so good to try if one might obtain that faver... I am S^{r}... y^{r} most oblidging frain & Humble sarvant
Thos Knowlton.'

Knowlton was most anxious to get the Double Nasturtium to grow in his garden. A mezzotint engraving of the plant with a pre-Linnaean name of 'Acriviola maxima odorata flore pleno' appeared in 1730 in the *Catalogus plantarum* published by a 'Society of gardeners'.(2) Previous to this in 1720 Hermann Boerhaave included 'Acriviola maxima odorata' in his *Index alter plantarum quae in Horto academico Lugduno–Batavo aluntur*, published at Leyden. Knowlton having gone to Holland in 1723 and 1726 may have brought a plant back with him after one of those visits. A native of Peru, *Tropaeolum majus* L. was introduced into Europe in 1684 and cultivated in the gardens of Count Beverning in Holland.(3)

(1) The Latin word pila (plural pilae) meaning anything round, a ball, or globe of material.
(2) The figure is not an accurate representation of *Tropaeolum majus*.
(3) Miller (P.), *The gardeners dictionary... The eighth edition,* etc., (1768).

In a postscript to his letter Knowlton wrote:

'... just am I a going to walling fen for y[e] pilla [*Cladophora aegagropila*] & Lentibulair [probably *Utricularia vulgaris* L., Greater Bladderwort][(1)] whch grows there as dos y[e] Glyzeraz [Knowlton may have intended to write Glyceria in which case he probably found *Glyceria maxima* (Hartman) Holmberg, Reed-grass, or possibly *Glyceria fluitans* (L.) R.Br., Flote-grass] ... I intend to Bring a q[ty] of y[m] a Longe with me so will put y[m] into y[e] watter where you see millitare azoides [*Stratiotes aloides* L., Water Soldier] att 8 in y[e] evening I just now come home from walling haveing Brought allmost one Bushell of y[e] pillas it can never by y[e] mottion of y[e] watter that forms them if so they would be found as well one sides & shallows as in y[e] deeps & Deepest watter ... I shall send you by the carer [carrier] y[e] Pilla some time at convenance & hope these will find you well &c when I was att whitbey I was Informed of a genteman that had bult a new stove att a place cald carlton not above 10 miles from thence, & y[e] genteman's name presswick a grat Lover but nither he nor his gardiner understands it & has both ben att London to make purchess.'

In a later letter Knowlton wrote again of this gentleman and his stove. In the eighteenth century when more and more new plants were being introduced from abroad a greatly increased number of greenhouses and stoves had to be built for exotic plants requiring heat. It will be remembered that soon after James Sherard went to Eltham he was busy in 1720 building a greenhouse and two stoves and when Knowlton went to Yorkshire Burlington promised to provide him with a stove at Londesborough. This, however, had not yet been built and one can imagine Knowlton's feeling of envy when he heard the news about Presswick and his gardener.

After completing his letter Knowlton added a note, dated 29 August 1729, and a list headed 'such plants I have observed to grow wild that are not very common'. He wrote:

'I here have coped over what you wanted but in hast so hope youl excuse faltee speling &c ... I am Resolve to buld a small stove to preserve some few things & what ever you think I have not shall be pleased [to receive] ... y[r] specimen of my honysuckle I should be glade to see & hope youl mak a stay in this country I desyne att making a must to come y[r] way

(1) The name given to this plant in Ray's *Synopsis*, 286, is 'Lentibularia'.

so as to have satisfaction or pleasure to see y^r collection I hope youl excuse haste being Hurred with workmen of all sorts & kinds maks me thus.'

The list cites the names and localities of most of the plants recorded by Knowlton in his letter of 12 January 1727/28 addressed to Patrick Blair,[(1)] with the addition of a number of new entries, including the following:

'... asparagras sativa [*Asparagus officinalis* L., Asparagus] grows plent in all ye meadow in Gurnsey as Likewise dose comone figes [*Ficus carica* L., Fig] Ledum palustre nostras arbuti flore [*Andromeda polifolia* L., Marsh Andromeda] grows att ye moses [mosses][(2)] att yould[(3)] in marshland plenty & in yorkshire Empetrum montanum [*Empetrum nigrum* L., Crowberry] in many places in yorkshire phalangium anglicum palustry [*Narthecium ossifragum* (L.) Hudson, Bog Asphodel] att yould mosses
oxycocus sive vaccine palustre [*Vaccinium oxycoccos* L., Cranberry] att yould mosses...
mistole [*Viscum album* L., Mistletoe] apon ye maple betwen doncaster & colingham [Collingham, two and a half miles south west of Wetherby on river Wharfe, both places in the West Riding of Yorkshire], & one [on] ye Aria Theophrastij [*Sorbus aria* (L.) Crantz, White Beam] one [on] ye edge of ye downs neerre pettworth Sussex
padus Theophrastij [*Prunus padus* L., Bird-cherry] grows plenty in ye hegds bettween bolton abby & arlington Craven in Yorkshire
Alnus vulgaris subconis ligulis membranaceis rubris donata this I found in 1728[(4)] & continues ye same this yeare 1729 & grows as you Ride from Everingham to Harsewell [Harswell, about two miles south east of Everingham in the East Riding] as soon as your over ye Bridge turn in ye gate one ye Right & just within ye 2^d Close down by ye wattersides grows this tree [Details of this species known by the common name of Scarlet Alder appear in Philip Miller's *Gardeners dictionary* (1731) with a note reading 'This sort was found in a meadow near Long-leet, [Longleat, two and a half miles west of Warminster, Wiltshire] by Mr. Brewer, some years since; but it is a matter of doubt, whether it is a distinct species, or some accidental

(1) *See* above, pp. 85–91.
(2) Moss: a dialect word used in Yorkshire, and other northern counties, for a bog, swamp, a peat bog.
(3) It is learnt from Camden (W.), *Britannia... The second edition,* etc., **2**, Col. 889 (1722) that Youle was 'nigh the meeting of Dun and Humber'. This is now known as Old Goole, situated near the confluence of the river Ouse with the Don.
(4) *See* also above, p. 100.

variety.'(1) It was probably *Alnus glutinosa* (L.) Gaertner cv. Rubrinervia].(2) Rhamnoides fructifeara salicis folio baccis leverter flavescentibus seu olester Germanicus [*Hippophaë rhamnoides* L., Sea Buckthorn] and some thousands of y^m^ as Likewyse ye eringos eringam marinum [*Eryngium maritimum* L., Sea Holly. Knowlton saw both these species between Lythe and Whitby]. pilla palustris [*Cladophora aegagropila*] about 2 fett [feet] & half & up in that meare next to Humber in walling fene [12 miles west of Hull]'.

TO SIR HANS SLOANE

Londesburght yorkshire Set^r^ ye 20 1729(3)
'Worthy

S^r^ I persum to send you the following sorts of specmens – Knowing y^r^ grate Love for all ye parts in naturall; history; more espeachaly the botanical study...

The first is specmen's of the alder [Scarlet Alder](4) found by Mr Brewer in ye west of England & Last year by my self att a placee call'd Harswell not farr from this placee & Likewise obeserved the same this year...

The other is as rair to me being a sort of morsse [moss] wch I call pilla aquatica nither can I find it in any author I have tho Wormius has one of ye sea which he calls pilla marina a conferva & seems to agree only he takes his to be formed a conferva by ye motion of ye watter into that shape; but [the specimen I send you]...I am well asured [is] of the sort to ye contrary; its to be had near this place in a mear called walling fen mear some 10000'd of em it struck me with suprising pleasure when I first see it...'

Sir Hans Sloane (1660–1753), born in Co. Down, lived most of his life in London and achieved equal eminence as physician, scientist and collector. He was intimately involved in the affairs of the Royal Society, and was its president from 1727 to 1741, and corresponded with many of the eminent scientists of the day. His collections, when purchased by the government in 1753, formed the basis of the British Museum.

The pilae which have already been mentioned by Knowlton in earlier letters were probably formed from masses of green algae, named by Linnaeus in 1753 *Conferva aegagropila* and now called *Cladophora aegagropila* (L.) Rabenh., which is rolled into balls by the water. This is the first known

(1) After the sixth edition (1752) of the *Gardeners dictionary*, the entry on the 'Scarlet Alder' is omitted.

(2) *See* Bean (W. J.), *Trees and shrubs hardy in the British Isles... Eighth edition*, etc., **1**, 276 (1970).

(3) Sl. MS. 4050 fol. 200–201 in British Library.

(4) *See* below, p. 155.

Sir Hans Sloane (1660–1753), eminent physician and scientist, whose collections formed the basis of the British Museum when they were purchased by the government in 1753.

record of the occurrence of these balls in the British Isles.[(1)] Knowlton in telling Sir Hans Sloane about finding what he called 'pilla aquatica' in Walling Fen, wrote 'Wormius has one'. This is a reference to Ole Worm, in Latin Wormius, a Danish doctor of medicine, born at Aarhus in 1588 and died at Copenhagen in 1654, whose volume in Latin entitled *Museum Wormianum* was published at Leyden in 1655. The present writer is indebted to Miss K. P. Kavanagh of the Department of Botany, British Museum (Natural History), for an English translation of a passage on page 139 of Worm's work, which reads:

'Pila stagnalis, a spherical structure, is called after the size and shape of the greater globular fruit of a palm tree, and because it was found in a swamp near Sora Saelandiae [Sorø, Zealand, Denmark, a town near which is a small lake]. It is composed of the tiniest hairs or tufts as it were, massed into the most exquisite roundness. While fresh it was observed to be green in colour, and herbaceous; now in the dried state the outer surface tends towards yellowness, while inside the tufts still appear green. The structure, when its size has been taken into account, is exceedingly light. For although in size it is comparable to a fist or the green globe of a palm tree yet it weighs no more than a single drachma. It seems to have developed thus from the constant thrust and rotation of water, out of mossy and grassy tufts of aquatic plants...'[(2)]

Knowlton gave Sloane details of another plant. He told him:

'I never see [it] att home nor abroad till 2 years since; nor do I find it in any catuloge; it being ye same as in the Hortus amst: 2 vol 159 calld thire Lathyrus distoplatyphyllus Hirsutus; mollis magno & perameno flore odoro but as that is a fine purple so on ye contrary this is a pure white ye uper Lipe onely tending to a Carnation collour; if y^r^ Hon^r^ has a mind to any seeds I shall have some ripe in a short time & will be att y^r^ sarvce; & glad should I be if I had any thing worthy y^r^ exceptance...'

The book mentioned was the *Pars altera* (1701) by Caspar Commelin of

(1) For further details concerning *Cladophora aegagropila See* Hoek (C. van den), *Revision of the European species of* Cladophora, 51–8 (1963) Leyden, and Newton (L. M.), A beach-ball mystery at Torbay, *Illus. London News* 96 (21 January 1950).

(2) For further details regarding the pilae formed in the swamp near Sorø *See* Acton (E.), On the structure and origin of 'Cladophora balls', *The New Phytologist,* **15**, 1–10 (1916).

Horti medici amstelodamensis rariorum . . . plantarum . . . descriptio et icones, etc. This fine folio volume consisting of plates and descriptions of plants was published at Amsterdam. On plate 80 is an illustration of the plant now called *Lathyrus odoratus* L., Sweet Pea, and a description of it is on pages 159–160.

Later in this letter Knowlton referred to his success when he was gardener to the Duke of Chandos at Cannons in raising Logwood trees (*Haematoxylum campechianum* L.). He named several northern plants any of which he would be glad of the opportunity to send to Sloane and told him:

'Aloe aquatica; s. mellitaria azoides [*Stratiotes aloides* L., Water Aloe, or Water Soldier] I fetched from wotten abby(1) & have y^m^ grow very will in severall Cannells in the parke & no dough would do ye same in Chelsey Garden in there watters. It is a most elegant plant worthy ye Gratist care . . . about 2 years ago ffound ye Christophorana [*Actaea spicata* L., Herb Christopher] in grate plenty between Hoverngham & Slingsby; Likewise the Rhamnoides fructifera salice folio baccis leverter flavesentibus [*Hippophaë rhamnoides* L., Sea Buckthorn] . . . between Lythe & Whitby as in ye Synopis [of John Ray, p. 445 (1724)] August . . . Last.

I have caused a figur to bee Drawn of the Chameapericlymenum [*Cornus suecica* L., Dwarf Cornel] for one of my men wch is a going Down into Scotland & have promised him a giney if hel take pains to sarch Chervett & pentlands Hill [i.e. Cheviot Hills in Northumberland and Pentland Hills in Scotland] where ye synops & Mr Wood says it grow's; upon condition he sends it to me I am very much mistaken if I did not see it in flower in ye Leyden Garden in ye year 1726 the flower white like ye sassamum [*Sesamum orientale* L., Sesame] & D^r^ [H.] Borhaave tould me it was that in his Index plantarum . . .'

Knowlton was quoting from the third edition, edited by J. J. Dillenius, of John Ray's *Synopsis* p. 261 (1724) where we read:

'Chamaepericlymenum . . . Dwarf Honey-suckle. In Northumbriae montibus Chevioticis dictis . . . (At Pentland-Hills five miles from Edinburgh plentifully; Mr. Wood.).'

Hermann Boerhaave (1668–1738) was the Professor of Medicine and Botany at Leyden and he also held the office of 'praefectus horti' of the

(1) Possibly Watton abbey near Great Driffield. *See* below p. 126.

Hermann Boerhaave (1668–1738), Professor of Medicine and Botany at Leyden, whom Knowlton met on his visit to Holland in 1726.

University botanic garden. On the above occasion Knowlton took the opportunity of presenting to the Professor a copy of Dr James Douglas's *Lilium sarniense: or a description of the Guernsay-lilly*. Knowlton told Sloane that Dr Douglas had given him half a dozen copies of the book to give away.

We shall find that Knowlton often assisted friends, who issued their works in parts, to get subscribers for their publications. One of these authors was the naturalist and artist Mark Catesby (1683–1749), who published *The natural history of Carolina, Florida and the Bahama Islands.* Although it is recorded that the first part appeared in 1730[(1)] it seems that it was in fact issued the previous year, as Knowlton told Sloane in the present letter of 20 September 1729:

'My old aqu[aint]ance M[r] Catesby has sent me down a part of his ffine History of Caralina haveing become a subscriber to it & hope shall procure him some more...'

Knowlton also informed Sloane that he had sent his 'worthy fraind D[r] Richardson some of the pilla's'; Richardson had never seen this species before and was much pleased with the gift. Finally, the gardener ended his letter:

'I...hope this will come safe to y[r] hands being sent up by y[e] York Carrer w[ch] sets out from york one tusday next & will be in basing hall street [London] the Thursday 7 nights following next & with submision am with y[e] gratest respect S[r] y[r] most honourd...obydant humble Serv[t] & a Lover of Gardening & Botany
Tho Knowlton'

A note in another hand records that this letter was 'Recd. Octob.'

Let us now consider some interesting details concerning Knowlton's discovery of *Cladophora aegagropila* in Walling Fen. The position of this flooded tract of land was described by Richard Gough in his edition of William Camden's *Britannia*, Vol. **3**, p. 98 (1789) as follows:

'Conferva aegagropila [*Cladophora aegagropila*], Globe conferva, or Moor Balls; in Walling fen moor, 12 miles west of Hull, a lake of 1 or 200 acres,

(1) *See* Stafleu (F. A.) and Cowan (R. S.), *Taxonomic literature,* etc., **1**, entry 1057 (1976).

on a large Common, subject to be over flowed by salt water from the Humber. This lake has been drained since these vegetable balls were described in Phil. Trans. XLVII for 1751 and 1752 p. 498 [1753].'

This description in the *Philosophical Transactions* was read at a meeting of the Royal Society held on 18 June 1752. It consists of an extract of a letter, dated 1 June of that year, from William Dixon F.R.S. of Loversall near Doncaster to the physician and naturalist William (later Sir William) Watson F.R.S., 'Concerning some vegetable balls; with remarks on them by... Watson.'

At the end of his description Dixon wrote:

'Whether they are particular to this place, I know not, having no description of them in my botanic library. To you [i.e. William Watson], who have so general an acquaintance in that branch of natural knowledge, they may prove old acquaintance.'

Regarding this passage Watson remarked:

'The vegetable here mention'd, and which I take the liberty of laying before you, I have never seen till now; neither have I been able to find it described in any of the botanic writers, whom I have consulted. The matter, of which it is composed, is that of a *conferva*; and should therefore have had a place under that genus in Dillenius's *Historia muscorum* 1741 [1742]. They are of a deep-green mossy colour, are hollow, of an irregularly spherical figure, and of different sizes, from an inch and half to three inches in diameter. They are cover'd with very short *villi* externally, and the thickness, from their external to their internal surface, is about a quarter of an inch; their texture is most compact the nearest to the surface. I should denominate them globose *conferva*.'

It is evident from this letter that neither Dixon nor Watson had any knowledge of the finding by Knowlton, many years earlier, of the pilae in Walling Fen. This was probably due to the fact that he never published a description of his discovery. He first wrote to Samuel Brewer about the green alga on 12 April 1728,[1] and it is probable that about the same time he told Dr Richardson about it. Subsequently, on 23 October 1729

(1) *See* above p. 98 and letter dated 24 August 1729.

Richardson sent to Sir Hans Sloane various subjects of natural history interest, together with a letter in which he said:[1]

> 'I have put up with the rest two Pilae Stagnales Worm: Mus: p. 139: found in a large Lake nigh Beverley Yorkshire, called Hornsea mear [*sic*] M. Tho. Knowlton... who dwels nigh this Lake, sent me these, & told me that there were thousands of them in the bottom of the water, which lay like Tennis Balls. Wormius tells us that these Balls are made of a Conferva &c. by the Agitation of the Water. I observ'd all those that I had seen were hollow.'

This letter was read at a meeting of the Royal Society held on 6 November 1729.[2] Shortly afterwards at a meeting held by the Society on 13 November Sloane read the above letter he had received from Knowlton, dated 20 September.[3] Neither letter was published in the Society's *Philosophical Transactions.* It is curious that Dillenius did not include the pilae in his *Historia muscorum* as he would have realized that they were cryptogams, but perhaps he was never told about them. By the time Dixon wrote his letter in June 1752 it seems likely that no one remembered anything about Knowlton's 'sort of moss' which he called 'pila aquatica'. Moreover, two of the recipients of Knowlton's letters namely, Brewer and Richardson, were dead, and Sloane, the third person Knowlton informed about his discovery, was a very old man. It was not until nine years after Knowlton's death that his discovery of the green alga was acknowledged in print when, in 1790, Richard Pulteney wrote in his *Historical and biographical Sketches of the progress of botany in England* vol. **2**. 239:

> 'Thomas Knowlton... merits notice in these memoirs, were it only to record his discovery of that singular production, the Globe Conferva, or Moor Balls (Conferva aegagropila Lin.); which he first found in Walling fen Mere...'

When Knowlton in 1728 discovered the 'Cladophora Balls' cryptogamic genera were very poorly understood. In Caspar Bauhin's *Pinax* (1623), a valuable work in which are ranged the names of plants used by the

(1) *See* copy of letter in the *Register Book of the Royal Society.*
(2) *See Journal Book of the Royal Society.*
(3) Copy in *Journal Book of the Royal Society.*

older botanical writers, the algae are recorded under *Muscus*, *Quercus*, *Fucus*, and *Corallina*. Some years later John Gerard's *Herball* (Johnson's edition of 1633) listed algae under *Lichen*, *Quercus*, *Fucus*, *Muscus*, *Corallina*, and *Conferva*, while John Ray in his *Synopsis* of 1690 and 1696 gave algae under *Corallina*, *Fucus*, *Muscus*, and *Lichen*. An outstanding exception to the widespread failure to recognize the importance of generic concepts was J. J. Dillenius who in 1724 published the most important edition of Ray's *Synopsis* and in 1742 was the author of *Historia muscorum*.

The study of the structure and reproduction of algae was closely connected with the development of the microscope. This is evident from the fact that the beginnings of rationalized genera were products of the very late eighteenth and early nineteenth centuries – a time which saw the gradual improvement in the quality of the lens and in the magnification of the microscope. Even in the last quarter of the eighteenth century algae were mostly designated as *Fucus*, *Corallina*, *Ulva* , and *Conferva*, except in a few works such as John Stackhouse's *Nereis britannica* of 1781. During the earlier years of the nineteenth century the groupings of algae into major Divisions on grounds of pigmentation also began to be appreciated.

TO DR RICHARD RICHARDSON

Londesburght Octr ye 29 1729(1)
'Worthy S^{r}

After my Long silance this wait one [on] you with my most Harty humble sarvice; for the faver of y^{rs} wch was Rec'd soon after the date; & most humbly beg pardon for not anserwing it sooner I have had the pleasure of hearing of y^{r} wellfair by Mr Brewr & S^{r} walter Hawksworth whom I have got to become a subscrber to Mr Catesbys History of Carolina wch I persum both you & the good Dr Stanope are Incoragers of; my self have become one some time since & Likewise have got him 3 more from his Ld-Ship who was very willing from my Recom̄endation... I Likewise shall procure him some more; when I wait upon Esqr willeby(2)... wch will not be Long he has sent for me this 10 days but our grate & good L^{d} with his Lady being heare could not as yett; I Likewise the other day had a Letter from Dr Chamber's who's very well.(3) I Longe Likewise to see you & y^{r} Deare garden but having a call to London this winter beside's a grate deal of Bussnes of his L^{d}Ships & with other gentlemen in this naiborhood... fear

(1) Richardson corres. MS. RAD. Trust. C.7.
(2) Regarding Willerby *See* Below, p. 148.
(3) Regarding Dr Chambers *See* above, p. 98.

shall not til Spring; but however have the satisfaction to tell you I have bult a small stove that will in my Imagination do any thing in Relation to ye most tender sorts so shall when I go to London procure a cargo as cofe [*Coffea arabica* L.] pineaple's [*Ananas comosus* (L.) Merr.] guagus [*Psidium guajava* L., Guava] pupas [*Carica papaya* L., Papaw] plantans [*Musa* cv., Plantain] wch I am Resolve to use my utermost skill in — I remember you was so good to tell me if any thing new offerd... you would transmit it to me Indeed things of that kind are allways shure to find the most harty Reception & in ye mean time am with humble sarvice to y^r^ good Lady & self y^r^ most obydunt Humble sarv^t^
Tho Knowlton.
PS You'l very much obldge me by Letting me know how Mr Brewer dos for I have not heard a Long time Indeed I sent him a Box of things & a letter by ye post but have not once heard since so was afraid he was not well...'

TO SAMUEL BREWER

Londsburght Novr ye 17 1729(1)
'Worthy S^r^

I had the pleasure of y^r^ Last week for wch I Return you my harty thanks for that & all other favers... w^ch^... [you] have so offten Confered apon me... Since I wote to you Last wee have had our good Lord & Lady down for some time this autome I am Likewye very glad to heare of y^r^ agreable Devertission one [on] fossell &c... as to ye Stars Stons [i.e. the stem ossicles of fossil crinoids] you mention att Bugthorpe [about six miles north west of Pocklington, in the East Riding] I was there when y^r^ Letter came att a gentlemans seat... [for] too or 3 days one [on] account of measuring his grownd for a new Disigne for a grarden & paddock & shall be that way againe within a month... so will take care of y^m^ as Likewise att Crack [or Cracke, about three miles north east of Howden, in the East Riding] having a fraind Liveing neare that plaice... who will take care of any thing I shall ask of him

I am extremely sorry to heare of y^r^ Late affliction of the Rhematism...

M^r^ fairchild's Death... was in all the London papars being a hamsome account of his past Life with feunerall &c well worth reading & is to Long otherwise should tran'scribe it for you; I have Likewise since been advertized of it both by Mr Baccon & Mr Catesbey the first haveing a few sorts of trees from the other...'

(1) Knowlton corres. British Museum (Natural History).

Stephen Bacon, nephew of Thomas Fairchild, inherited his uncle's garden stock at Hoxton. This nursery garden was noted for its collection of North American exotics which owed much to Mark Catesby, who sent Fairchild seeds and plants from Virginia and Carolina.

To return to Knowlton's letter, he told Brewer:

'I thak you for yr kind offer of Double nastertum's I have it my self from London about a month since by my man Jno...I sent up one [on] some Bussness one [on] his L^{d}Ship's...I have Likewise bult a new stove & shall this winter go to London & then shall have anna's [*Ananas*, Pineapple] & other curiosity from thence so hope to entertaine a fraind [Brewer?] with their sight when he comes to see me I have Likewise sent Sr Han's Slone severall things wch came to his hands very well & had most kind Letter from him & his generious gift of too voll when ever I please to send for [it, i.e. Sloane's *Natural history of Jamaica*, etc. 2 vol. 1707, 1725]'

Knowlton continued by thanking Brewer for an offer of some plants, apparently intended for the stove. These he did not accept, partly because of the risk involved in transporting them in winter, and partly because until he went to London and ordered some pots he had nothing suitable to put them in.

Earlier, it will be remembered, Knowlton asked Brewer if he had a copy of *Plantae per Galliam, Hispaniam et Italiam observatae* with plates by Jacques Barrelier and text by Antoine de Jussieu.(1) Brewer must have replied in the affirmative as Knowlton now wrote:

'Barrelleos Iecons are cheep in my humble opinion I could wish I could come by y^{m} as easey...

I hope...shall have ye Pleasure to heare of y^{r} welfeare & shall be glad to see you after my Return & would meett you att york to atend you one [on] for this plaice where I should be glad to see you in ye spring I should be glad to heare how Mr Martain Decads gos one [on] & what father he Intends to publish.'

John Martyn, Professor of Botany at Cambridge from 1732 to 1762, published his *Historia plantarum rariorum* in five decades from 1728 to 1737, but the work did not meet with sufficient encouragement for it to be continued after the fifth.

(1) *See* above p. 105.

In a postscript Knowlton gave fulsome praise regarding Sir Hans Sloane:

'Sr Hans Slon's generosity is so extensively grate that scarce any prince in Europe can parrelle [parallel] it & Indeed notheing can more signife his Grate Love then his such truly generious spirite how wonderfull Diferant is this from what I have heard from others who hardly could give a good word for him...'

TO SAMUEL BREWER

Londesburght feb^{r} ye 8th 1729/30(1)
'S^{r}
I am not yet gone fro London & haveing this opertunity by our worthy prist [priest] send you these to Informe you that since my Last(2) I have been att Bugthorpe & Lepinton [Leppington] where I could not find ye astroides [i.e. what Knowlton previously called 'star stones', the stem ossicles of fossil crinoids](3) but got an old woman very quick that way who pick[ed] up these 2 or 3 I hear send you but att this season they are not very easly to be found the time for it is ye sumer after a hasty shower of rain wch washe's y^{m} so clean that they are easly found I have spook to ye same old woman who I am shure well take care to procure me a quanty of y^{m} There found some times with in 2 mils of this place [Londesborough] Likewise but raire; but att Lepinton in ye town street in Clays are very plenty & there call y^{m} Lepinton + as crosses &c'

It is interesting to note that Bugthorpe, Leppington (two miles north of Bugthorpe), and Londesborourgh (about eleven miles south east of Bugthorpe), all in the East Riding, are all on clay of the same geological age — lower lias clay. Knowlton continued:

I have Likewise writen into Craven for ye troches or entroches [trochites or entrochites, wheel-like joints of the stem of fossil crinoids] a wonderful ode [odd] stone or matter & hope to Rec'd them some time in ye springe I should be glad if you can spare a few of those sorts of ston's that have ye Impresions of filix [fern] kinds one [on] y^{m} & should be glad if youl be so

(1) Knowlton corres. British Museum (Natural History).
(2) *See* above, letter dated 17 November 1729.
(3) *ibid.*

good to send a sample of too or 3 sorts of y^{m} by the bearer... the star's ston's are well figured & Described in Mr Chamber's Diccinary'

On 12 January 1727/28 Knowlton told Blair that he had been making woods and water. Three months later on 12 April 1728 he recounted to Brewer that he had made several pieces of water and that he was stocking them with fish. In the present letter of 8 February 1729/30 he informed his friend of further excavations to be filled with water:

'I think Dr Richdson in his Lettar tells me he'l come over with you to this place in ye spring I have just as many Cannills as ye D^{r} in our parke finishd & full of watter & am now about other 2 so shall have a fine chaine of y^{m}...'

The word 'cannill' used by Knowlton, usually spelt canal, is defined in the *Oxford English Dictionary* as a long and narrow piece of water for the ornamentation of a garden or park, the word apparently deriving directly from the seventeenth century French 'canal'. Such artificial stretches of water were made in the Royal garden at St. James's, at Bramham in Yorkshire, at Euston in Norfolk, and in many other gardens and parks of the seventeenth and eighteenth centuries.[(1)] The so called 'canals' at Londesborough, however, were of every conceivable size and shape, resembling ponds and lakes rather than canals.

Knowlton continued his letter, thus:

'I hope my Deare Double nasturtum's [*Tropaeolum majus*] well & will be perseved over ye winter & that youl have the pleasure of seeing it in perfection in y^{r} garden wch I hope to see some time next spring... when I shall gett Leausure to Inspect y^{r} curiositys wch will be so much to my satisfaction... shall tomorrow seat out fro London... I have sent you some trochites wch are wonderfulle prity & can if you want y^{m} send a Large Quantity of em the Starr Stone only one but shall have many of y^{m} in my Return... I should be Glad to know how Dr Stanopes garden & plants with y^{r} own & Dr Richardsons Dos... I am with kind Respect y^{r} most obytant Humble sarvant
Tho Knowlton
NB if you should have any od seeds to spair I shall hartyly accknowledge faver... the person that went Chevett hills & pentland Hill for Dear Chamepericlymenū [*Cornus suecica* L.] y^{t} Returnd but with out it.'

(1) *See* Cecil (Hon. Mrs Evelyn), *A history of gardening... Third... edition,* 185, 223, 226 and illustration facing p. 176 (1910).

It will be remembered that Knowlton promised one of his men who was going north a guinea if he would search for the Dwarf Cornel in the Cheviot Hills in Northumberland and the Pentland Hills, South of Edinburgh(1). Knowlton had noted that the plant was recorded in Ray's *Synopsis* ed. 3, 261 (1724), as growing in both these places. Both areas were extensive and Knowlton's man was unsuccessful in his search. Furthermore, he found that the plant was not represented in the Edinburgh Physic garden, neither was it growing there earlier, in 1716, when George Preston published his Catalogue of the plants growing in that garden.(2) In the Physic garden Knowlton's man met 'a gentleman whose name was Allstone'. This was presumably Charles Alson who in 1716 received the appointments in Edinburgh of King's Botanist, Keeper of the Royal Garden, and Regius Professor of Botany.

TO SAMUEL BREWER

Londesburght March y[e] 28 1730(3)
'S[r]
I Recd Both y[r] favers for w[ch] I acknowledge my selfe much oblidged and am very glad to heare of y[e] Return of y[e] worthy D[r] Richdson & hope hel soon be well... I am extreemly will pleased to heare of your wellfaire; as all frainds are in the South... y[e] Grate S[r] Hans Slone that prince of Botanist[s] whom I waited [on] to Return thanks for favers Recd... showed y[e] utermost satisfaction for those small favers I had sent to hime the watter aloe I cared [carried] up alonge with me & several other things... I hope y[r] plants are all very well & more particularly y[e] Double nasturtum...'

The writer continued by telling his friend that once again he had been unsuccessful in growing the Double Nasturtium. Furthermore, he had the ill luck of losing his Pineapples, Plantains, Guavas, etc. Never discouraged Knowlton wrote to Brewer:

'I shall when over Bege Leave to Robe you of a bit of my beloved nasturtium w[ch] I call my own being first Brought into y[e] this kingdom by me...'(4)

(1) *See* above, p. 112.
(2) Entitled, *Catalogus omnium plantarum, quas in seminario medicinae dicato transtulit Georgius Prestonus,* etc., (1716) Edinburgh.
(3) Knowlton corres. British Museum (Natural History).
(4) A similiar remark was made by Knowlton in his letter dated 24 August 1729, see p. 106.

In the following remark Knowlton was referring to a recent publication:

'I have ye compeneys Book of gardening it costs 1^{ll} 11^{s} 6^{d} a very Ridiculus price I see full of nothing but faults Lik the Company...'

On 17 February 1730, the month before this letter was dated, the *Catalogus plantarum... By a Society of gardeners* appeared. Was this the book to which Knowlton was referring? This Society, composed of a number of the principal nurserymen and florists in the vicinity of London, published the *Catalogus* in an effort to establish a botanical nomenclature. Considerable confusion was caused at that time both to the trade and to the public by the attribution of different names to the same plant. The preface lists the names of a number of owners of English gardens, including James Sherard of Eltham, who introduced into this country 'many new kinds of plants, flowers, trees and fruits'. Only the first part of the work on trees and shrubs was published and among the 21 plates of plants are several figures of flowers prepared for a later part, but these figures have no descriptive text.[(1)] One of these figures, as stated earlier, depicts the 'Double Nasturtium'.[(2)] One can imagine Knowlton's disappointment, perhaps even annoyance, at finding no details regarding this plant he thought of so highly.

It is interesting to note how carefully Knowlton kept up with news regarding works to be published or recently published. After referring to the Company's Book he wrote:

'...my fraind M^{r} [Philip] Miller is just now publishing proposall[s] for a Book in gardening being about 1 pd & 5 shiling...I shall have the proposall[s] in a short time & will send or Bring y^{m} over in order for you to assist wch will be a very good thing & worthy every ones purchasse Likewise my fraind m^{r} switere [Switzer] is one [on] another wch will in it[s] way be extradinary published by subscription to wch I shall Desire yr assistance...'

Knowlton was correct in thinking that the first of these works would be worth purchasing. Philip Miller (1691–1771), for many years the eminent

(1) For further details regarding this publication *See* Henrey (B.), *British botanical and horticultural literature before 1800*, etc., **2**, 212–13 (1975).

(2) *See* above, p. 106.

John Miller (1715–c. 1790), the German artist and engraver who did much of his work in England; this portrait has in the past erroneously been thought to represent Philip Miller, of whom no likeness is known to exist.

gardener to the Society of Apothecaries at Chelsea, published the first edition of his famous *Gardeners dictionary* on 13 April 1731. According to *The Monthly Catalogue* for February 1730 the work was to be published by subscription and the price to subscribers was to be £1. 5s. 0d. in sheets. The popularity of Miller's book is evident from the number of editions in which it was published, the eighth edition being issued in 1768.

Stephen Switzer the noted seedsman, garden designer, and horticultural writer is generally considered to be the editor of a periodical publication stated on the title-page to be 'By a private Society of husbandmen and planters'. It was presumably to this project that Knowlton wanted Brewer to subscribe. The work appeared in monthly parts, commencing in May 1733,(1) and when complete it was issued under the title of *The practical husbandman and planter* in two octavo volumes, dated 1733 and 1734, respectively.

It will be remembered that when Knowlton wrote to Sir Hans Sloane on 20 September 1729 he had received the first part of Mark Catesby's *Natural history of Carolina.* The second part appeared in 1730, and in the present letter of 28 March 1730, he told his friend:

'Mr Catersby has published another seat [set] our L^{d} had three of y^{m} when I was in town & was pleasd to say to me it was ye best done he ever see any thing of that Kind I was glad to heare his L^{d}Shipe was so well pleased I hope it will be for m^{r} Catesby['s] advantage his L^{d}Shipe Intended to see him att fairchilds with ye Aloe & would have had me stay to wait one [on] him... but Bussness att home prest me to tell him so & hope his Ld-Shipe would not take it Il'

The Earl of Burlington would have been able to see Catesby at Fairchild's because the naturalist worked for a time at the Hoxton nursery, first under Fairchild and later under his successor Stephen Bacon.(2)

Knowlton collected books on many other subjects besides botany and gardening. He shared with his first wife an interest in ornithology and we find him writing to Brewer:

'M^{r} albine is... publisheing an history english & forren Birds 80 plats [plates] being already engraved I havee proposalls & specimen's by me

(1) *See The Monthly Catalogue* for May 1733.

(2) *See* Henrey (B.), *British botanical and horticultural literature before 1800,* etc., **2**, 275–76 (1975).

& have enter'd myself one & promisd to gaine some others it is well done & is humble proposed att 4 pd & 4 shilling in quarto good paper & in naturall colours if you can procure any for hime it will be a faver acknowledged...'

This work when completed appeared in three quarto volumes during the period 1731 to 1738. It was published by Eleazar Albin (fl. 1713–59), an artist of natural history subjects, under the title of *A natural history of birds*, etc. The many plates coloured by hand are accompanied by descriptive text.

Knowlton closes his letter by thanking his friend for some seeds.

TO SAMUEL BREWER

May ye 12 1730[1]
Londesburght
'S^r

I had both y^r favers for wch I ought to have acknowleged before now but a grat Hurry with all sorts of traidsmen of ever[y] kind, & mateorales [materials] to provide & y^e other Bussness in gardening &c. wch are no small things have Imployd my head & hands [and] are the resons of this omition...I have thought of seeing y^r feasts if posable but our grat & good Lord talks now of comeing down to see y^e alterations att this place wch have been many since $Sept^r$ Last is y^e Reson of y^e unsertanty.

I wrote to y^e worthy D^{rs} gardener [i.e. Richardson's gardener] to bespeek me a Dos. of spads [spades]... will write to Henery Smith nail man of y^r town to bring y^m with a load or 2 of nails, as to y^e water aloe [*Stratiotes aloides,* Water Soldier] & whatever else of specimen's you want [that] I can save you I shall [let you have] with pleasure to y^e utermost of my capasity... with... grate hast... from S^r y^r very Humble $Sarv^t$
Thos knowlton'

A study of eighteenth-century gardening accounts reveals the large quantity of nails, of all sizes, used at this time.

In a postscript Knowlton again referred to the Scarlet Alder and *Stratiotes aloides*. In an earlier letter he had referred to the latter as growing at 'Wotten abby,'[2] but now he gives the locality more precisely, writing:

(1) Knowlton corres. British Museum (Natural History).
(2) *See* above, p. 112.

Eleazar Albin (fl. 1713–1759), naturalist and painter, wrote and illustrated a series of books on insects, spiders and birds. His A natural history of birds *was published in 1731–1738.*

'Millitaria azoides plenty betwen Hull & Beverly one [on] Both sids ye way in ye Dicks[1] & att a fuling mill near Wotton Aby.'

Knowlton possibly meant Watton Abbey, spelling this name phonetically as 'Wotton aby'. The abbey was a Gilbertine house,[2] about five miles south west of Great Driffield. The nearby fulling mill was employed for washing, scouring, or fulling cloth.

(1) Dick or ditch (*O.E.D.*).
(2) *See* Sheahan (J. J.) and Whellan (T.), *History and topography of the ... East Riding*, **3**, 516–19 (1856).

TO SAMUEL BREWER

[No address or date](1)
'S^{r}
I wrote to you about 3 days since & after that have Rec^{d} y^{r} other Letter wch comes to tell the spads [spades] are Ready I have written to Mr Henery Smith nail man in Bradford for a quantity of nails by whom I would have ye spads come wch I desire youl be so good to tell him... Likewise... beg youl be so kind to send a few of those sorts of ston's wch have ye Impresions of plants one [on] y^{m}... by ye same Henery Smith... Il take care to Return ye mony [for the spades] or else come over myself when... I hope to have ye pleasure of seeing you & ye worthy D^{r} to whome I bege youl be so kind to give my Humble sarvce to him & his good Lady...
Thos knowlton.
about a forthnight since I was in Holderness the most estermost part of it & Returnd by Hull but found no veritity [variety]; Eryngo's [*Eryngium maritimum,* Sea Holly] was very plenty ye watter aloe [*Stratiotes aloides*] I know of in many spots but all Londesburght shall have 100's of y^{m} they multiply so I shall take care to observe y^{r} directions in every particulare when I come wch I cannot Determin [as] ye care of workmen of all sorts & many of y^{m}, is a grate tey [tie]... if you have a plant of ye Double nasturtium & could spare one I should aceknowlege the faver... I am &c &c y^{rs} againe & againe.'

TO SAMUEL BREWER

June ye 16 1730(2)
'S^{r}
I had the faver of yrs for wch I Return my Humble thanks & have herein sent you Mr Millers proposalls wch I hope youl endeavor to promote... I hope to procure him about 20 [subscribers?] &c
as to the fosells you so kindly offr'd will be by no means desired if with ye smalist dificulty... procured... my self haveing but very Little judgement that way... the Astragalus Il take care of wch is just now in flower...
I have my Deare nasturtium done againe & hope now will be Safe... all frainds att London are well & my self & whole family... I am so excedinly engaiged with workmen of every Kind can by no means come over till some time in ougust besids our grat & best of Lords is comeing Done one [on] purpos to se the alteratans I have maide wch are more within these six

(1) Knowlton corres. British Museum (Natural History).
(2) Knowlton corres. British Museum (Natural History).

months than all ye 4 years... before. I am y^{rs} with Respect
Tho Knowlton
NB Mr Priswick att Carlton 10 mils or thereabouts from Whitby has bult a new stove as I have been Inform'd & perhaps may want something to furnish [it with] I have been tould he's a pasonate Lover I know he has often writ to some of my frainds...'

In a letter to Brewer of 24 August 1729 Knowlton mentioned the building of this stove and also the fact that the owner and his gardener had visited London to make some purchases.(1) It seems possible that this 'pasonate Lover' was 'Codrington Prissick of Carlton in the county of York, Esq.' who appears as a subscriber to Philip Miller's *Gardeners dictionary* of 1731. Of the many Carltons in Yorkshire, the only place with an important residence was the parish of Carlton, or Carleton, about three and a half miles south south west of Stokesley in Cleveland, North Riding. In the reign of James I the manor of Carlton was granted to Charles Bruce of Kinloss, ancestor to the Earl of Aylesbury, from whom it came by purchase to the Prissicks. The manor house of the Lords of the Manor had good gardens and stood on the west of the village.(2)

Knowlton kept up to date with works published abroad as well as in England. He tells Brewer about a new Russian publication:

'... we have a new Book of botany come over from petersburght [St Petersburg] a midling 4to & about a Ginaiy [guinea] & is chefly observations of ye plant[s] about Constantinople &c.'

This work was by the German botanist Joannes Christianus Buxbaum (1693–1730) who went to Constantinople to study the plants. As a result of his researches *Plantarum minus cognitarum centuria I* [–*V*] *complectens plantas circa Byzantium & in Oriente observatas* was published at St Petersburg. The first and second centuries appeared in 1728, the third, fourth, and fifth in 1729, 1733, and 1740, respectively. Buxbaum died during the printing of this important work which includes some new and many rare plants.

(1) *See* above, p. 107.
(2) Graves (J.), *The history and antiquities of Cleveland in the North Riding, of the County of York*, 159–60 (1808).

TO SAMUEL BREWER

Londesburght July y[e] 8 1730(1)
'S[r]

I Recd y[r] faver for [ch] I am very much obl'dged & in Return send a few of such small things I have & are as follows

N 1 asteroides [Star stones] &c
2 the thunder bolt
[Thunderbolt, locally applied to various stones, fossils, or mineral concretions, formerly or vulgarly supposed to be thunderbolts]
3 Cockle shells out of y[e] Bodeys of Larg Stone att Birdsall [about three miles south east of Malton in the East Riding] &c
4 the watter alos [*Stratiotes aloides* L., Water Soldier] 6 plants for the worthy D[r] Richardson & y[r] self
5 specimens of y[e] astragalus [possibly *Astragalus danicus* Retz., Purple Milkvetch]
6 2 verburnum's [possibly *Viburnum lantana* L., Wayfaring Tree or *Viburnum opulus* L., Guelder Rose] the one for D[r] Richdson
7 & 2 plants of y[e] purple amarhthoides [i.e. Amaranthoides, *Gomphrena globosa* L., native of tropical America] y[e] one for y[e] D[r] Likewise.
8 Eryngium marrinum [*Eryngium maritimum* L., Sea Holly] from whitby
9 plantago paniculus sparsis [a form of *Plantago major* L.,Great Plantain] &c found in yorkshire 2 years since
10 a sensative &c. [possibly *Mimosa sensitiva* L., native of tropical America]
11 & the Birds peper seu Capsicum siliq minima Rubr [*Capsicum frutescens* L., Bird Pepper, probably native of tropical America](2)
12 primula veris flore Rubro [*Primula farinosa* L., Bird's-eye Primrose] from fountain abby
13 Senna Hexaphyla [*Cassia bicapsularis* L., native of Tropical America] from Bengall & M[r] [Peter] Collinson.
14 Helleborine flore albo ger: [ger: = Johnson's edition (1633) of Gerard's *Herbal*, *Cephalanthera damasonium* (Miller) Druce, White Helleborine] wild att fountain abby
15 Double flowering stramonium [possibly a double form of *Datura metel* L., Horn of Plenty, native of India] &c

w[ch] are what I though you perhaps had not, nor yett y[r] neubours for w[ch] Reson I hope the'll be y[e] more wellcome ... sorroy ... I have not y[e] Ladies sliper [*Cypripedium calceolus* L.] but one plant & not flowerd ... this year.'

(1) Knowlton corres. British Museum (Natural History).

(2) The classification of the genus is very confused (Purseglove (J. W.), *Tropical crops. Dicotyledons,* 524 (1968)).

Even in Knowlton's day the beautiful, but now extremely rare orchid *Cypripedium calceolus* was an uncommon plant. It is listed in William Camden's *Britannia* Vol. 3, 99 (1789), under 'Rare plants found in Yorkshire' as growing in Helks Wood, near Ingleton and in the neighbourhood of Kilnsey, both places in West Riding.

Together with this letter Knowlton sent to Brewer proposals for printing Eleazar Albin's *Natural history of birds* in the hope that his friend would get subscribers to the work. He also acknowledged receiving the spades, adding that he wanted to know if Smith of Bierley wished him to pay the nail man. If he did, then he would pay the man provided he brought a receipt with him witnessed by Mr George Pickergill, whose hand he knew. Otherwise payment would have to wait until he himself visited Bierley which probably would be 'sometime in harvest', if Lord Burlington was not then in residence.

TO SAMUEL BREWER

Londesburght August ye 22^{d} 1730(1)
'S^{r}

y^{r} Rec^{d} y^{r} faver this day by m^{r} farer & yesterday y^{r} Box by tom Long ye carier & am very sorry I have given you so much trouble to no purpose having ye mortification to find all dead as well I might after so long confinement; I bege... youl not send me any more by that Dam'd fellow... as to y^{r} Box I would send it [back] but the way these things come is so dificult that it will not be worth while without I had somewhat vallable to put up in it & then perhaps all perish as it is my Il soul the Double nasturtium I have had twice this Sumer & still want it for it never grew with me so shall forbare to send for it any more & besides have so very much Bussness apon my hands to think of that in short I have but very Little time... to spare...

I am extremly sorry thos few things I sent should have ye same dam'd delay for I knew ye fellow whent derickly for Bradford & not by carier as my Box came from one [carrier] to ye other wch I knew must pass thro' thre or 4 of y^{m} [carriers] before it could come to me but now sure hope youl excuse this Long & hasty scrible... with much Respect Sr Yrs
Tho knowlton'

When Knowlton said that he knew the goods he sent to Brewer 'whent derickly for Bradford' it is likely that he sent them by a man making the journey privately on horseback. The box that came to him from his friend,

(1) Knowlton corres. British Museum (Natural History).

however, was sent by carrier. From remarks made by the gardener in several of his letters it is evident that it was not uncommon for goods despatched by carrier on cross-country journeys to be delayed and sometimes to be damaged as well. It is probable that the box sent by Brewer was taken to the carrier at Bradford and from there transferred in turn by two or three carriers to Market Weighton, by way of Bradford to Leeds, Leeds to York (or perhaps directly from Bradford to York) and from York to Market Weighton whence it would be fetched by someone at Londesborough. As it is unlikely that the arrival of one carrier and the departure of the next synchronized, the journey would often have been slow. Moreover, at this time, and later, the state of the roads was very bad and carriers could seldom exceed a speed of four miles per hour with their broad-wheeled, heavy, canvas-topped wagons. Furthermore, carriers were sometimes accused of carelessness in looking after the goods under their charge.(1)

TO SAMUEL BREWER

Londesburght June ye 28 1733(2)
'S^r^

Yesterday I Recd y^r^ faver & am very glad to heare of your wellfeare & that you Intend to p[rint] y^r^ observations &c wch will be very agreeable to me & hope to every Body else that hath a hart ye same way'

It may be assumed that the physician and naturalist Richard Pulteney was referring to this project when he wrote:(3) 'Brewer . . . meditated, and nearly finished, a work which was to have borne the title of "The botanical guide", but it never appeared'. We shall find that Knowlton, after learning of this undertaking, was most anxious that it should succeed and he often asked for news concerning its progress.

Earlier it was noted that Dr J. J. Dillenius, the German botanist who in 1721 was persuaded by William Sherard to come to England, began in 1724 to make drawings and descriptions of the new plants in James Sherard's garden at Eltham.(4) The resulting two large folio volumes entitled *Hortus elthamensis*, although bearing the date 1732, were apparently

(1) *See* Jackman (W. T.), *The development of transportation in modern England, second edition revised* (1962).

(2) Knowlton corres. British Museum (Natural History).

(3) *See Historical and biographical sketches of the progress of botany in England*, etc., **2**, 190 (1790).

(4) *See* above, p. 41.

not published until the following year, when they were listed in the *Monthly Catalogue* of books for March 1733.[1] Subsequently, Knowlton received a copy of the work:

> '...att Last I have the D.rs H. Eltham: & find several plants figuer'd very well in severall authors before & wonder what usse that repitition can be...to Lovers...but indeed ye generall part are Intierly new therefore it is the more welcome to me & every othear ye same way Inclynde; the gratest part of those plants was Draon by ye D^r^ when I Lived att Eltham...I ...wonder he did not put it into English as well as Lattin wch would have maide it sell better besides might be done att ye same charge as ye H. A — mensis is.'

'Lovers' of botany and gardening must have been delighted with this production with its detailed descriptions, and illustrations depicting plant introductions, outstanding for their accuracy. Knowlton regretted that the text was only in Latin, a fact that must have disappointed many persons not well versed in that language. The chief buyers of such a sumptuous work would have been the owners of gardens and estates with fine libraries who would have found the Latin text no problem. Latin was also universally understood among persons of good education abroad. However, books were increasingly being published in the vernacular and the work Knowlton gave as an example was issued at Amsterdam with the text in Latin and Dutch, in parallel columns. It was Jan Commelin's *Horti medici Amstelaedamensis rariorum...plantarum...descriptio et icones*, etc., published in two volumes in 1687–1701. The second volume of this posthumous work was by Jan's nephew Caspar Commelin. If the *Hortus elthamensis* had been similarly produced with text in the vernacular as well as in Latin sales might indeed have been higher but so would the purchase price, due to heavier printing costs resulting perhaps in fewer copies being sold rather than more.

Knowlton continued by telling Brewer that he would be very glad to see him, the sooner the better as he was expecting Burlington on a visit, and when the Earl arrived he would have little spare time. Regarding the visit, he added:

> '...when you are heare you may be welcome to any Books I have or

(1) *See The Monthly Magazine 1732–66 monthly catalogues,* **2**, 167 (March 1733). Republished by Gregg Press in 1966.

plants... & if you give me notuce will send a man & horse to tedcaster [Tadcaster, nine miles south west of York] to mett you... all Loses I believe may be repaired with some aditionall plants when you are heare.'

By 29 October 1729 Knowlton had fulfilled his wish and built a stove.(1) Subsequently he became increasingly interested in exotic plants. This is evident from the following passage:

'... the plants that came from Lancastershire are very few... &... most very small bitts or cutings... that it was next to an Impsoability to preserve y[m] & severall of y[e] plants so bad could not with all my art make y[m] grow what I have... that are a Live or worth mentioning are as follows:

The Large mellon thistle [*Melocactus communis* Link & Otto, Turk's Cap, native of Jamaica]...
plant of marrum [*Teucrium marum* L., Cat Thyme, native of the western Mediterranean]...
Cusion [cushion] aloe [*Haworthia retusa* (L.) Haw., native of South Africa]...
Kettmia chinesis folio Digitato R [probably *Hibiscus mutabilis* L., commonly called China Rose, native of China](2)...
Dimont fioccoides [Diamond Ficoides, *Mesembryanthemum crystallinum* L., Ice Plant, native of South Africa]...
opuntia spinis valid[issimis]: [*Opuntia tuna* (L.) Miller, native of the West Indies]'

With the same consignment of plants Knowlton had been sent a Plantain, *Musa* cv., and 'a poore weeke coffee tree', *Coffea arabica* L., but both these plants died. However, he was not very upset at these losses as subsequently he received another Plantain 'and 3 or 4 fine cutings of y[e] coffee tree'. He was determined to grow tropical fruits successfully, and he told Brewer, '... if I Live... shall very soon compleat a garden to my wish.'

(1) *See* above, p. 107.

(2) Knowlton was probably referring to 'Hibiscus sinensis foliis cordato-quinquangularis obsolete serratis, caule arboreo (Syn. Ketmia sinensis, fructu subrotundo), China Ketmia with a roundish fruit, commonly called China Rose.' *See* Miller (P.), *The gardeners dictionary... The eighth edition* (1768). This species had earlier been called *Hibiscus mutabilis, See* Linnaeus, *Species plantarum,* 694 (1753). The species that Linnaeus named *Hibiscus Rosa sinensis,* China Rose Hibiscus, was called in Knowlton's day, 'Hibiscus foliis ovatis acuminatis serratis glabris, caule arboreo. *See Species plantarum,* 694.

Of one of the figures in Dillenius's work Knowlton remarked to Brewer:

'... in y[e] H[ortus] E[lthamensis]... that monstrous plant you gave Cons[ul William Sherard] a specmen of w[ch]... he would not belive was any real one but now I see it ther figured under ye name of Corona &c with a complement to ye Dorhum Garden'

This plant named by Linnaeus *Fritillaria regia* and now called *Eucomis regia* (L.) L'Hér. is illustrated by Dillenius in the *Hortus elthamensis* on plates 92–93. It is a native of South Africa. The owner of the garden Mr Blathwayt (spelt Blatwait by Knowlton and Blaithwait by Dillenius), was the son of the politician William Blathwayt (1649?–1717). He was Lord of the Manor of Derham or Dyrham, about five miles from Chipping Sodbury, in Gloucestershire. A park adjoined the gardens which were originally laid out by Le Nôtre.(1) Knowlton had recently been offered plants from Blathwayt's garden, as is evident from the following passage:

'Mr Mesenger was att Mr Blatwait's & M[r] [John] power [the gardener](2) tould him any thing in that garden was at my sarvces if I would send for it ye trechelium [possibly *Campanula trachelium* L., Bats in the Belfry] was Dryd up in y[e] Garden wall y[e] weather being so hot, but all those plants you writ for I shall have from thence... M[r] Mesenger sent me the Lunaria minor [*Botrychium lunaria* (L.) Sw., Moonwort] about a week since y[e] finest & Largest I ever see he is Latly come from Bath & tell me y[e] navele wort [*Umbilicus rupestris* (Salisb.) Dandy, Pennywort, Navelwort] he see one [on] a Bridge there in plenty.'

The 'Mr Mesenger' was presumably the gentleman of the same name who lived at Fountains Abbey.(3)

Knowlton's interest in botanical publications is again evident in the following passage:

'M[r] Catesby has published his fifth part [of *The natural history of Carolina, Florida and the Bahama Islands*] with a preafece & is gone over to Holande he wrot to me to Know if he could be of any sarvice to me there. M[r] Swit-

(1) Rudder (S.), *A new history of Gloucestershire,* 427–29 (1779) and the D.N.B.

(2) Mr John Power is given as gardener to Blathwayt, in Dillwyn (L. W.), *Hortus collinsonianus,* etc., 10 (1848).

(3) *See* above, p. 98.

zer has pubshed his monthly treatais of husbandry &c [i.e. *The practical husbandman and planter*] as to amaranthus you may have what specmens you please if you think y^{m} of any worth — I know there is one of martains...'

In the above passage Knowlton was referring to the figure in John Martyn's *Historia plantarum rariorum*, entitled 'Amaranthus sinensis foliis variis, panicula elegantur plumosa', the Purple Amaranth, called by Linnaeus *Amaranthus cruentus*. It is a native of China.

Finally in a postscript, dated 'July ye 3^{d}', Knowlton seems to show some anxiety regarding Brewer's proposed visit clashing with Burlington's arrival. He told him:

'I am dayley in exspectation of our good Lord but can not be sertain when hel be hear... I am allways att this time of ye yeare extreemly bussey & much more so exspectinge ye noble l^{d} &c'

TO SAMUEL BREWER

august ye 27 1736[1]
'Deare S^{r}

I Recd the faver of y^{rs} dated ye 16th of august & am very sorrey you should be disopinted in ye stramonium [possibly *Datura metel* L., Horn of Plenty] wch I... belive will flower yett I Know it was a fault to pack ym up in to small a Box but had not any but would have been as much to big & was afraid of makeing ye same chargable... ye Reson I did not send you citerrons [*Citrus medica* L., Citron] ye stocks was 4 fett high & most of ye oranges Lemons of ye Low size was gone ye others I was afraid would have been to Large & to cumbersome to you besides ye carrdge would be exspensive in Large casses [Among oranges and lemons grown in gardens at this time the most common were *Citrus aurantium* L., Sour or Seville Orange, *Citrus sinensis* (L.) Osbeck, Sweet Orange, and *Citrus limon* (L.) Burm. f., Lemon] I have buded some of ye scarse Kinds wch if take well; can easely send one of each sort another yeare you must Know those I Bought ware all forain stocks Buded in genoa 3 or 4 years before they came I have 60 od Large ones yett to Dispose of tho I exspect Mr Aisalby will buy a parsell he haveing orderd his gardener to wright to me.'

(1) Knowlton corres. British Museum (Natural History).

In writing of the Citron tree Philip Miller recorded:[(1)]

'There are several varieties of this fruit, with which the English gardens have been supplied from Genoa, where is the great nursery for the several parts of Europe for this sort, as also orange and lemon-trees; and the gardeners who cultivate them there, are as fond of introducing a new variety to their collection, as the nursery-men in England are of a new pear, apple, peach, etc.'

The 'M[r] Aisalby', whom Knowlton expected would buy some of his trees, would have been the statesman John Aislabe (1670–1742) who from 1714–18 was Chancellor of the Exchequer. He was the owner of an estate at Studley, about two miles from Ripon, in the West Riding of Yorkshire, celebrated for its fine grounds through which flowed the river Skell. He also owned Hackfall, a beautiful but less extensive property about five and a half miles north west of Ripon through which ran the river Ure.[(2)]

It is evident from his correspondence that Knowlton kept himself up-to-date with local and current topics of interest by reading the newspapers. In the above letter he wrote of an account that appeared in the *Country Journal: or the Craftsman* of 21 August 1736 concerning a riot that had taken place at Lord Petre's property in Essex. An outstanding botanist and horticulturist, Robert James Petre (1713–42), eighth Baron Petre of Writtle, was the owner of Thorndon Hall, near Brentwood in Essex, with a garden famous for its many rare trees and shrubs. The account of the riot that was published in *The Craftsman* is given below; such disturbances were not unusual among labourers during the eighteenth century:

'The Right Hon. the Lord Petre having for some time past employ'd a considerable number of labourers to make canals and gravel-walks in his gardens at his seat in Essex, on Monday last several of the country assembled in a tumultuous and riotous manner, and assaulted the said labourers, suspecting them to be all Irishmen so that a battle ensued, in which heads were broke, and many bruises received; and his Lordship was forced to order his fire-arms to be brought out before the rioters would withdraw.'

It may be noted that during the eighteenth century the Irish were drawn

(1) Miller (P.), *The gardeners dictionary... The eighth edition* (1768), *See* under *Citrus.*
(2) *See* Camden (W.), *Britannia... Enlarged by...* R. *Gough,* etc., **3**, 57 (1789).

to England by economic considerations in increasing numbers. Their strength and endurance enabled them to get employment, usually at relatively low rates of pay, in such occupations as digging, heaving, and carrying. However throughout the century and especially in times of depression, animosity grew against the immigrant Irish and tumults often resulted.[(1)]

At the time of writing Knowlton had plants of:

'... ye male papa [*Carica papaya* L., Papaw] & ye aburus.'

The last named plant may be what is now known as *Abrus precatorius* L., Crab's Eyes or Red Bead Vine. The seed is scarlet with a black spot. It is a native of the West Indies and was introduced into this country in 1680. Tournefort in his *Institutiones rei herbariae*, 393 (1719) gave it the name of 'Orobus americanus, fructu coccineo, nigrâ maculâ notato.' Linnaeus in the *Hortus Cliffortianus* (1737) [1738], placed the plant in the genus *Abrus*.

The gardener also told his friend that he had, 'ye china aster [*Callistephus chinensis* (L.) Nees, native of China]... a most gloryus flower' and that 'Mr Telford', a member of the firm of nurserymen of that name at York, was cultivating 'ye virg: Honeysuckle, [*Lonicera sempervirens* L., the Trumpet Honey-suckle, native of Virginia] but being tender they dye yearely'.

Knowlton, it will be remembered, wrote on the 28 June 1733 that he was very glad that Brewer intended to print his botanical 'observations'. In the present letter he mentioned the project again:

'I hope now you have begun y^r^ Botanick Book youl hartyly aply your self to it... ye many fine plants you have Discover'd will be a grate adition to ye Synopes [by John Ray]... I hartly wish you health Longe Life & a trew & sincere pease of mind... y^r^ most obydant Humble Sarvent
Thos Knowlton
pray mak my complements exceptaible to ye noble D^r^ &c & Lady'

In a postscript he added:

'I have grat plenty of young coffe trees [*Coffea arabica* L.] if you want one... Likewise papaws [*Carica papaya* L., commonly known as Papaya,

(1) *See* Ashton (T. S.), *An economic history of England: the 18th century*, 10, 32n, 222, 227 (1955).

Papaw or Paw-paw, native of the tropics] wch I propigate by cuting I have aloe genensi [*Aloe guineensis,* now called *Sansevieria guineensis* (L.) Willd., African Bowstring Hemp, native of West Africa] but have Lost ye aloe zeylanica [*Sansevieria zeylanica* (L.) Willd., native of Ceylon] & seaverall more fine things... to Long to Repeat y^r^ Beautyfull Helianthemums are whit & pale read collour'd flowers wch are very prity...'

Knowlton was probably referring to *Helianthemum nummularium* (L.) Miller, Rock Rose, of which there are many garden forms in a great variety of colours.

TO SAMUEL BREWER

Londesburght Octr ye 1st 1736[(1)]
'Deare S^r^

I was Glad to heare in y^r^ Last that you think you shall finish y^r^ Bot: Guide by the next spring... the following is a List of such Plants... wch I have not if you can furnish me with some of y^m^ perhaps I may by ye generiosity of m^r^ Blackburn mak some others.'

The list is a long one. John Blackburn (1690–1786), from whom Knowlton hoped to obtain some of the species he wanted, lived at Orford, near Warrington, Lancashire. He is said to have built the first hot-house in the north of England[(2)] and his garden was famous for its fine collections of exotic and indigenous plants. Some of the plants Knowlton included in his list were the following:

'absinthium arb. [*Artemisia arborescens* L., Tree Wormwood, native of southern Europe]
adatoda [*Justicia adhatoda* L., Malabar Nut, native of the East Indies]...
chamalea tricocus [*Cneorum tricoccum* L., Widow-wail, native of the western Mediterranean region]...
convolulus canariensis [*Convolvulus canariensis* L., Canary Bindweed, native of the Canary Islands]...
cucurbitifera arb: [*Crescentia cujete* L., Calabash Tree, native of Jamaica and Barbados]...

(1) Knowlton corres. British Museum (Natural History).
(2) *Extracts from the literary and scientific correspondence of* Richard Richardson, etc. [Edited by Dawson Turner], xxx (1935).

John Blackburn (1690–1786), who grew many exotics in his garden at Orford Hall near Warrington in Lancashire, was reputed to have been the first to build a hot-house in the north of England.

Digitalis canariensis [*Isoplexis canariensis* (L.) Loudon, native of the Canary Islands]...

Dorycnum [*Dorycnium pentaphyllum* Scop., native of central and southern Europe]

Lentiscus [*Pistacia lentiscus* L., Common Mastic Tree, native of the Mediterranean region]...
Lotus st. Jacobii [*Lotus jacobaeus* L., native of the Cape Verde Islands]
Thlaspidium arb: Licoii follio [*Iberis semperflorens* L., commonly called Tree Candytuft, native of Sicily and the west coast of Italy]...'

Knowlton told his friend that if, by the following spring, he could supply him with some of the plants he needed, he intended to go to Bierley to fetch them. At the same time he hoped to see Brewer's Botanical guide ready for the press. He warned that if the work was to contain plates the engravers would make a great deal of trouble for him as well as expense.

During the eighteenth century there was, as we have seen, a constant exchange of plants among enthusiastic botanists and horticulturists and great interest was shown with regard to the successful cultivation of exotic fruits. Thus Knowlton said to Brewer:

'I sent you by Smith what Plants I had that you wrot for my collection is pryty considerable considering I have noby [nobody] to assist one in it I see in ye St James Evening post Sept ye 2^{d} to 4th[1] the plantain or mussa in fruit att Sr Joseph alife att Harrow one [on] ye Hill so we are just behind ye Holanders in ye same perfection[2] I have a very Large plants & so tall ye House will not hould em but not any signe of a flower.'

The name here mentioned is a mis-spelling on Knowlton's part for Ayloffe. The rate book for the parish of Harrow covering the period 1728–66[3] lists Sir Joseph Ayloffe as paying a rate from 1734 to 1737 inclusive, for a property in Sudbury, a hamlet due south of Harrow on the Hill, but no details are given concerning it. Peter Collinson in a memorandum, published in *Hortus collinsonianus*, p.34 (1848), a work edited by L. W. Dillwyn, wrote:

'Plantain, or Musa, first fruited in England at Sir Joseph Ayloffe's at Harrow on the Hill; another at Mr Blackburn's and Lord Petre's.'

(1) No copy for this date appears to be extant.
(2) Possibly a reference to the plant that flowered in 1736 in the garden of Gerard George Clifford, at Hartecamp, near Leyden, Holland.
(3) Ref. DR03/B2/1 Greater London Record Office (Middlesex Records).

Anxious as always to hear about recent publications with a natural history interest, Knowlton told Brewer that he would be glad to know if he had Kaempfer's *History of Japan*. This work, written in High-Dutch by Engelbert Kaempfer (1651–1716), was translated into English by Sir Hans Sloane's librarian, Dr Jean Gaspard Scheuchzer, and published in two volumes in 1727 in London. It includes an account of Japan's metals, minerals, trees, plants, animals, birds, and fishes. Knowlton asked Brewer's opinion about this publication and also about a work being published in parts:

'...if you have Kemphers [*History of Japan*] pray Lett me Know how hes aproved of by ye Larned wether he Describes any plants & figurs [in] ye same if well done & if you Know ye prise of y^m^ because I have some mind to bey y^m^ if you think y^m^ worth my purchessing pray Lett me Know how many of Jno Martains Decads...are of plants...I supose [it] is a very pompous work.'

In this passage Knowlton again refers to John Martyn's *Historia plantarum rariorum*, a work he first wrote to Brewer about in a letter dated 17 November 1729,(1) and which ceased publication in 1737. It is a sumptuous publication with many of the plates made from drawings by the Dutch flower painter Jacob van Huysum, and it is of particular interest as it contains some of the earliest examples of colour-printing from a single plate. These plates were executed by Elisha Kirkall, an engraver noted for his mezzotint engravings.

In a postscript Knowlton told Brewer:

'I have not had a Lett^r^ from London from any aquantance a grat while I shall soon [write] to Mr Collinson & Catesby wch will then transm[it] to me all ye news of ye curiouse things wch Like y^r^ self I am allways well pleasd to Read & heare of.'

Peter Collinson (1694–1768) was noted for his interest in natural history. A merchant with a business as a haberdasher and mercer in London, at the Sign of the Red Lion, Gracechurch Street, Collinson was a great promoter of botany and horticulture. Through his trade with the American colonies he became acquainted with persons from Carolina, Virginia, Maryland,

(1) *See* above, p. 119.

Peter Collinson (1694–1768), a London haberdasher and mercer, whose trade with the American colonies enabled him to grow many North American plants in his garden at Peckham and later at Mill Hill.

Pennsylvania, and New England, and he sought their aid in sending him new and rare plants and seeds. At the time that Knowlton wrote his letter, Collinson had a noted garden at his cottage at Peckham in Surrey, full of all kinds of rare species, especially from America. Later, in 1749, he transplanted his treasures from Peckham to a new property, Ridgeway House in Mill Hill, and this garden became equally renowned.

Knowlton mentioned another famous publication, namely, Elizabeth Blackwell's *A curious herbal* (1737 [–39?]), but before giving his opinion concerning the work let us consider what Pulteney thought of the project.

Richard Pulteney (1730–1801), a physician with a practice at Blandford in Dorset, was responsible for various publications on botany and medicine, including his *Historical and biographical sketches of the progress of botany in England* (1790). In this book, which is still one of the most valuable and original publications on the subject, the author said (**2**, 251): 'It is a singular fact, that physic is indebted for the most complete set of figures of the medicinal plants, to the genius and industry of a lady'. This was Elizabeth Blackwell (d. 1758) who received praise for her book from a number of other authors, including the eminent botanist William Curtis.(1) The work contains 500 plates with figures of plants, accompanied by 125 plates with engraved text giving short descriptions of the plants, their place of growth, time of flowering, uses, and common names in several different languages. Knowlton, however, did not approve of Mrs Blackwell's work and the reason is not far to seek. The eighteenth century saw a vast increase in the number of new and rare plants introduced into our gardens, and Knowlton was anxious to have publications illustrating these new species. Of *A curious herbal*, he scathingly remarked:

> 'Mrs Blackwell [s] Herball y^{t} comes out by N^{rs} is a very silly stupid . . . thing I wonder att any person atempting . . . ye most cōmon Plants of all that have been . . . done as well [before] . . . I cannot see what usse that can be to Botany unlesse she minutely figured & Described all ye currouse new plants y^{t} are yearely Interdosed [introduced] . . . then I say there would be some what to be Learn'd but as it is; tis onely repeaiting over ye same that has been as well done by Gerard and Parkingson & many others old & venerable Authors . . . '

It might seem that to a person requiring illustrations of plants used in medicine the small woodcuts found in the works of the early herbalists,

(1) *See* Henrey (B.), *British botanical and horticultural literature before 1800*, **2**, 30 (1975).

including *The herball or generall historie of plantes* (1597) by John Gerard, and the *Paradisus* (1629) and *Theatrum botanicum* (1640) by John Parkinson, are in no way comparable with the clearly drawn, etched, and engraved folio sized plates, coloured by hand, in the two fine volumes of *A curious herbal* by Elizabeth Blackwell, but Knowlton was more interested in getting illustrations of the more recently introduced species.

TO SAMUEL BREWER

Octr ye 22^{d} 1736(1)
'Worthy S^{r}
by ye same post I Recd y^{rs} I had a Lett^{r} from D^{r} R[ichardson]. y^{r} naibor... about his new stove & a request for any thing I could spare to furnish ye same... I have answerd with... a List of what he may be welcome to... I have advissed... him to send his gardener over for y^{m} so if he dos I have desired hel ordur him to call... for those few plants wch you can spare without Robing y^{r} self...'

Among the plants Knowlton required from Brewer was 'anagalus moneli', a pretty blue Pimpernel, *Anagallis monelli* L., native of south-west Europe. At the same time Knowlton wrote that he himself had some of the plants that Brewer wanted. These included:

'aster aticus [*Aster amellus* L., native of Europe]
asphodelus flore albo candy [Candia (Crete)] as[p]erum [possibly *Asphodelus fistulosus* L., Onion Asphodel, native of the Mediterranean region and south-west Europe]...
Syringa ye comon onely [*Philadelphus coronarius* L., Mock-orange, native of a few places in Central and Eastern Europe]...
ye roses ye musk [*Rosa moschata* J. Herm.]
ye yellow [possibly *Rosa foetida* J. Herm., the Austrian Briar, also known as *Rosa lutea* Mill.]
ye velvett [*Rosa gallica* L. cv. Violacea]
ye cinamond [*Rosa cinnamomea* L.].'

We have seen how eighteenth century botanists and horticulturists anxious to build up their collections sought the help of friends who had

(1) Knowlton corres. British Museum (Natural History).

plants to spare. The alternative was to buy plants from the trade but this could be expensive, as is evident from the following passage written in a letter of 20 May 1731 by John Blackburn of Orford to Richard Richardson:(1)

'I have of late built some conveniences for keeping of green-house and stove plants, which have in some measure answer'd my expectation; and I have apply'd to some of the gardeners about London for plants of several kinds; for which they are so extravagant in their demands, that I find I can't fill my stoves and greenhouses, without being at great expense; and I take the freedome to write to you, to desire you will direct me which way to apply myself, and from whom I may be Supply'd on the most reasonable terms.'

It appears from Knowlton's letter that Brewer had asked him about a plant he called 'arundo variegato', now called *Arundo donax* L. var. *versicolor* (Miller) Stokes. Of this Reed-grass, commonly called the Giant Reed, with white striped blades, Knowlton wrote, 'I have never seen [it] in England its com̄on in the curriouse garden in Holand it is a very fine strip[e].' It will be remembered that when Knowlton visited Holland in May 1726 he met the Professor of Medicine and Botany at Leyden, Dr Hermann Boerhaave.(2) In Boerhaave's *Index alter plantarum quae in Horto academico Lugduno-Batavo aluntur*, published in 1720, *Arundo donax* is listed as growing in the Leyden botanic garden. It is native probably in the Mediterranean region, south-west Asia, and perhaps elsewhere.(3)

Knowlton told his friend, referring to Lord Petre on whose estate the previously mentioned riot had taken place:

'I am very much oblidge to you for y^r^ entertaining account of L^d^ petreas stoves & prodigious n^r^ of plants &c he is seartainly on of ye noblest spirits in this Kingdome.'

Lord Petre died of smallpox on 2 July 1742, at the early age of 29, and

(1) *Extracts from the literary and scientific correspondence of Richard Richardson,* etc. [Edited by Dawson Turner], 309 (1935).

(2) *See* above, p. 112.

(3) It is of interest to note that the plant was listed as growing in the Oxford Botanic Garden in the *Catalogus plantarum Horti medici oxoniensis* of 1648. *See* Pt. 1, p. 6, Pt. 2, p. 36. It was omitted however, from the 1658 edition of the work.

the naturalist Peter Collinson wrote of this peer's death that it has 'been the greatest loss that botany or gardening ever felt in this island'.[1] Later, on 4 April 1746 he gave some details regarding Petre's garden at Thorndon, Essex, in a letter addressed to Richard Richardson the younger:[2]

'As I am on a visit to Lady Petre's, I there have leasure to answer your's. Since our dear lord's death, my lady is so respectfull to his memory that the stoves and nurserys are kept in good order... The great stove is the most extraordinary sight in the world. All the plants are of such magnitude; and the novelty of the appearance strikes every one with pleasure. The trellices all round are cover'd with all species of Passion-flowers, which run up near thirty feet high: the creeping, great-flowering Cereus blows annually, with such quantities of flowers that surprises every one with their beauty, and at the same time perfume the house with their scent. There is variety of Cereus that have carried up their perpendicular heads to the very top; Papaws [*Carica papaya* L.]; Plantains [*Musa* cv.]; several sorts of Palms; Dragon-Trees [*Dracaena draco* L.]; Ficus Malabaricus [*Ficus religiosa* L.]; Rosa Chinensis [probably *Hibiscus mutabilis* L., commonly called China Rose[3]]; Acacias; Bamboos; besides a great variety of species raised from New-Spain seeds,[4] that I don't know. Many have flower'd and fruited: the Papaws hang now full of fruit and blossoms; and there is a most delightful show of several sorts of white Lilio-Narcissi now in flower.

The lesser stove, sixty feet long and twenty wide, is full of a vast variety of all species of tender exotics; and the nursery abounds with rare, hardy exotics, a great variety of Pines from North America not found in our gardens, Red Cedars and White Cedars, the great Magnolia and the small Magnolia flower every year in great plenty, Sassafras [*Sassafras albidum* (Nutt.) Nees], Sweet Gum or Liquid-amber trees [*Liquidambar styraciflua* L.], Yew-leafed Firr, Balm of Gilead [*Cedronella canariensis* (L.) Webb & Berth.] and Spruce-Firrs from New England and Newfoundland; with a great variety of all other hardy American trees and shrubs, in plenty.

To see his plantations now arrived to such perfection, the variety is so great that one would think one's self transported to some other country; American and European productions are so blended together.'

(1) Smith (Sir J. E.), *A selection of the correspondence of Linnaeus*, etc., **1**, 9 (1821).

(2) *Extracts from the literary and scientific correspondence of Richard Richardson*, etc. [Edited by D. Turner], 392–93 (1835).

(3) *See* above p. 134 footnote. There was no species called *Rosa sinensis* at this time.

(4) 'New-Spain, Vice-royalty of (1535–1821). Spanish colonial territory comprising, at its greatest extent, Mexico, the Caribbean islands, Central America north of Panama, the coast of Venezuela, Florida, and the present south western United States' (*See The new encyclopaedia Britannica,* 15 edition (1978)).

Knowlton, liking to keep up with the latest news concerning the gardens of his horticultural friends, wrote and asked Brewer:

'I should be Glad to Know how long D^{r} R[ichardson's] new stove is & where it stands & wether there is a tan bead in ye same it was a secreatt to me till m^{r} Blackburns Gard^{r} a few Days ago saide how he had been over att Bierly about ye same; I am Informd Mr Blackburn has a fine cinamond tree [*Cinnamomum verum* J. S. Presl] & Likewise some seeds of ye new tea tree wch a Relation of his Brought for [from] China.'

Richardson's stove and tan bed would naturally have been of interest to Knowlton.(1) Of even greater interest would have been the news that Blackburn had received seeds of the Tea tree. Great efforts were made by various botanists and gardeners in this country during the eighteenth century to grow this plant, but without much success. It was not until 1771 that the first Tea plant, *Camellia sinensis* (L.) O. Kuntze, flowered in England, at Syon House in Middlesex, belonging to Hugh Percy, first Duke of Northumberland.(2)

Knowlton's letter ends with the following postscript:

'I have not been from Home this severall weeks nor shall I next week so ye D^{r} had better send as soon as he can... because I imagine I shall be sent for to Mr Duncombs... very soon they I know are Drawing to conclusion of what I had... seat out... Likewise m^{r} willoughby where I have promised for a long time to go as soon as I was well recovered of my Late illness...'

Thomas Duncombe (d. 1746) was the owner of Duncombe Park, overlooking Rievaulx Abbey in the North Riding of Yorkshire, and it may be presumed that 'M^{r} willoughby' was the Honble Thomas Willoughby (1694–1742), son of Thomas, first Lord Middleton, and grandson of the seventeenth century naturalist, Francis Willughby, F.R.S. Willoughby (as the spelling was later changed to) was in contact with Knowlton as early as October 1729,(3) and it is of interest to note that it was in that year that Willoughby became the owner of Birdsall Hall, a few miles south east of North Malton, in the East Riding. Willoughby's wife, Elizabeth, was the daughter and heiress of Thomas Sotheby who at his death left Birdsall Hall

(1) *See* above, p. 81.
(2) *See* Lettsom (J. C.), *The natural history of the tea-tree,* etc., vi–vii, plate (1772).
(3) *See* above, p. 117.

to her and her husband. It was probably then that Willoughby sought the help of Knowlton in planning his new garden. Further evidence of his interest in horticulture is found in Phillip Miller's *Gardeners dictionary* of 1731, where his name appears in the list of subscribers.

TO DR RICHARD RICHARDSON

Londesburght oct^{r} ye 31 1736(1)
'S^{r}

This morning I Recd y^{r} faver, & have been in exspectation of y^{r} gardener for ye things wch he might safly att this season carrey... before him in a Bundle well paked with straw under him with straw apone ye pumell [pommel] of ye sadle & carrey y^{m} very easely as I have Done formerly many much Larger & a grater distance... the weather is so mild & Like to be that I take it thir cannot be any danger, if you have a Inclynnation for y^{m} & y^{r} stove warm enough to strick y^{m} so should be glad youl send as soon as you please, ye reson of my presing is I am so full of pine apple plants now y^{t} I am oblidge to mak room by turning out & Keeping but one of a sort of each I have of those things & Indeed am unwilling to thro y^{m} out being so many years in raising of y^{m}.'

In this letter Knowlton voiced his opinion concerning Linnaeus's sexual system of classification. At the beginning of the eighteenth century John Ray (1627–1705) in England, Joseph Pitton de Tournefort (1656–1708) in France, and August Quirinus Rivinus (1652–1723) in Germany, were the three principal systematic authors, and the system of each of these three botanists had its adherents. This resulted in much confusion and controversy. Ray's method of classification attributed importance chiefly to the fruit, Tournefort's basis of classification was flower-form, while Rivinus's views were based largely on the form of the corolla. The Swedish naturalist Carl Linnaeus (1707–78) decided to try and improve this state of affairs and put forward his 'sexual system' of classification which he introduced to the learned world in the first editions, issued in 1735 and 1737, respectively, of the *Systema naturae* and *Genera plantarum*. Linnaeus divided his system into 24 classes based essentially on the number of stamens, ending with the *Cryptogamia* whose reproductive habits were not well understood. Each class he subdivided into orders according to the number of pistils. The orders in turn were divided into genera and the

(1) Richardson Corres. MS. Rad. Trust. C. 9.

genera into species. The system of Linnaeus was artificial but it was simple and although at first its reception was cool in England, British botanists gradually became aware that it facilitated the study of botany, until by the second half of the eighteenth century most published systematic works were arranged in accordance with it. Thomas Knowlton was among those who disliked Linnaeus's 'sexual system' of classification, regarding which he wrote to Richardson:

'...as for Linaeus new method; 'tis Impossible for him or any other man whatsoer to go beyond a Ray, a Turnafort &c who have chose ye flower & fruit... [as to a single part of the flower it][1] is; in my opinion altogether whimsicall & Ridiculus...'

Knowlton continued by referring to a publication by Linnaeus entitled *Musa Cliffortiana florens Hartecampi*. He wrote:

'...now you mention his musa Clifordiana pray what is there more y^{n} complement to ye grate m^{r} Cliford & to tell us it has flowerd in his garden...'

In 1736 Linnaeus saw a Plantain-tree flowering under glass in the garden at Hartecamp, between Haarlem and Leyden, belonging to Gerard George Clifford (1685–1760), a banker and a Director of the Netherlands East India Company. This inspired him to write the above treatise. Besides describing the plant and illustrating it with two plates he also gave the nomenclature, systematic position, generic and specific characters, habitat and phenological details, uses, and literary references of *Musa*. Cheesman in his work *Classification of the Bananas* has pointed out that the work 'has much historical interest as a compendium of the "pre-Linnean" literature'.[2] 'Musa Cliffortiana' is identified with the plant later named by Linnaeus *Musa paradisiaca* in *Species plantarum*, 1043 (1753).

Knowlton showed himself to be a good practical botanist and gardener with an excellent knowledge of plants, both indigenous and exotic. Moreover, he was always anxious to acquire illustrated books with

(1) The words 'as to a single part of the flower it' have been taken from the transcription of this passage given by Dawson Turner in *Extracts from the literary and scientific correspondence of Richard Richardson*, 349 (1835). The words in the original are not clear.

(2) Cheesman (E. E.), 'Classification of the Bananas,' *Kew Bull.*, No. 2, 147 (1948).

descriptive text. However, works such as *Musa Cliffortiana florens Hartecampi,* he considered were 'maide for pompe to fill a Library & more for outward show than reale use'.

Before closing his letter to Richardson, Knowlton informed the doctor:

'... when y^r^ sarvant comes as I said before if I can spare any thing you want I shall not Denie him... I beg my compliments atend y^r^ good family & Mr Brewer & am S^r^ y^r^ most obydant Humble sarv^t^
Tho. Knowlton.'

TO DR RICHARD RICHARDSON

9 ber [November] ye 6 1736(1)
'S^r^

Ye under writen are the plants with n^bers^ &c you have a mind to

1–2 Large papas [*Carica papaya* L., Papaw, native probably of the American tropics]
2–1 Large flower fenel [possibly *Ferula communis* L., Common Giant Fennel, native of Mediterranean region]
3–2 solanum's
4–1 Guaver [*Psidium guajava* L., Guava, native of West Indies]
5–3 Tamarindus [*Tamarindus indica* L., Tamarind, native probably of Africa]
6–1 accasia cornigera [*Acacia cornigera* (L.) Willd., Bullhorn, native of Mexico and Costa Rica]
7–1 —— aldina
8–1 —— flore albo &c. Houston [possibly *Calliandra houstoni* (Miller) Standley, native of Mexico]
9–1 atriplex arb: Bengalensis [*Atriplex hortensis* L., Orache]
10–1 abutulon Indicum [*Abutilon indicum* (L.) Sweet, Indian Mallow]
11–1 Plantain [*Musa* cv.]
12–1 mahogony [*Swietenia mahagoni* (L.) Jacq., Mahogany Tree, native of West Indies, Central America, etc.]
13–1 hermania Gorss. [*Hermannia grossularifolia* L.,(2) native of South Africa]

(1) Richardson corres. MS. Rad. Trust. C.9. '9ber' i.e. The ninth month (November) of the Roman, or Julian Calendar, which was used in England till 1752. By this calendar the year began in March.

(2) Regarding this plant, Dr Verdoorn, Director of the Botanical Research Institute, Pretoria, has written to the present writer (in *epist.* 24.xii.79), as follows: 'The name *Hermannia grossularifolia* L. is correct for the species with the inflated capsule... The correct name for *Mahernia heterophylla* (Cavanilles (A. J.), *Dissertaio botanica,* **6**, 324, pl. 178 (1788)) is *Hermannia heterophylla* (Cav.) Thunb., a species quite distinct from

14–1 pigion pea [*Cajanus cajan* (L.) Huth, Pigeon Pea, native habitat unknown, widely cultivated in tropical countries]
15–1 flos passions [*Passiflora* L., Passion Flower]
16–1 ceruseu Quadrangularis [*Acanthocereus tetragonus* (L.) Hummelinck, Torch Thistle, native of Central America]
17–1 ficus Indicus Lattisimus [*Opuntia ficus-indica* (L.) Miller, Indian Fig, native habitat unknown, extensively cultivated in tropical and subtropical countries]
...I am...S^{r}...y^{r} most obydant Humble sarvant
Thos Knowlton.'

TO SAMUEL BREWER

'July ye 5th 1737(1)
I Recd y^{r} faver under cover for wch I paid D^{l} [double] postage wch to...give one [on] purpos to ye postmasters I allway think much of when it may as well be perVented; Josak Likewise Recd one in ye same maner from his father att ye same time wch was onely 2 quarter of sheets but supose they do not Know otherways or to be sure they would not put him to so much charge wch maid him grumble very much as indeed wel he might...to pay 6^{d} instead of 3^{d}...'

At that time letters were charged according to the number of sheets. By the Act of 1711 the rate was 3d. for a single letter (i.e. a single sheet, whether large or small) going 80 miles or less, and 6d. for a double letter.(2) Few people prepaid postage because to do so was taken as an insult implying the receiver could not afford to pay. Probably five per cent of mail outside London was prepaid.(3) Furthermore it was considered safer if the post office charged on delivery of the letter.

During the years 1737 and 1738 Knowlton, when writing to Brewer, mentioned Josak a number of times, spelling his name in the following ways, Josak, Jaseck, Josek, Jasek, and Josack. On one occasion he referred

Hermannia grossularifolia L. with the capsule not inflated. Unfortunately, Candolle (A.P. de) *Prodromus,* etc., **1**, 496 (1824), under *Mahernia heterophylla* gives as synonym *Hermannia grossularifolia* L. This is an error. The description does not include a capsule. In all possibility [Knowlton] was sending the true *Hermannia grossularifolia* L., the species with the inflated capsule and not de Candolle's concept which has a normal capsule and is a distinct species now named *Hermannia heterophylla* (Cav.) Thunb.'.

(1) Knowlton corres. British Museum (Natural History).
(2) Robinson (H.), *Britain's Post Office,* etc., 15, 53–4 (1953).
(3) Willcocks (R. M.) in *epist.* 29.xi.77. Willcocks is author of *England's postal history to 1840* (1975).

to him as 'Josek Roberts... my chefe man to ye stove'.(1) In the same period a person called Isack or Isaca worked for Knowlton and in one letter Knowlton wrote of him as 'florist Isack'.(2) A third man who worked for Knowlton in October 1737 is mentioned only by the abbreviation 'Jno' and in a letter dated 3 October 1738 Knowlton informed Brewer, 'M^r baine... is Bulding a stove for pine apples... in Lincolonshire where my man Jno is gone to Live'.(3)

After complaining at having to pay double postage Knowlton thanked Brewer for an offer of plants, but he told him that he was in no hurry for them. Moreover, he did not want his friend to deprive himself of any species he did not have plenty of 'nor yett', he added, 'to send any of L^d pettreas [Lord Petre's] favers; wch I have no Inclynation to Rob you of nor anything else...'

'Nor anything else' appears to refer to some cloth Brewer wished to give Knowlton for his son Charles,(4) a gift the gardener refused for the following reason:

> 'Charels shall be contended with my old close [clothes] mead up for him & goold for any child – it is Redeculus to fuse [fuss] & fill thir heads as half ye world do... over clothern it is what I have Know notion of nor shall he have y^t Indulgance from me while I Live what ever he may after death that will not then consern me so give you 10000 of thanks for y^r so grate a kindness in cloth wch by no means I would... Rob you of'

In this letter Knowlton offered to replace some plants that Brewer had lost and at the same time expressed sympathy with his friend regarding the absence of rain:

> 'The plants you have Lost I will replenish you again when ever you plase as ye pomgranites [*Punica granatum* L., Pomegranate] Egyptium & hornd accasea [*Acacia nilotica* (L.) Willd. ex Del., Gum Arabic; and *Acacia cornigera* (L.) Willd., Bullhorn, respectively] & perhaps ye cotten trees [*Gossypium arboreum* L., Tree Cotton] & Likwise you may have ye Colacaisa [*Colocasia esculenta* (L.) Schott, Coco-yam] again... I am very sorrey to heare your Grat want of watt^r. its a grat complaint in many places...'

(1) *See* below, p. 172.
(2) *See* below, p. 178.
(3) *See* below, p. 179.
(4) It will be remembered that at one time Brewer was connected with the woollen trade. *See* above, p. 92.

It appears from published records that the whole of the year 1737 was one of average rainfall in Yorkshire,(1) but from the above remark in the present letter of 5 July 1737, and from a reference made to the season 'being so continual Dry & scorching,' in a later letter of 9 October of the same year, it is evident that some areas at least were short of water. In Knowlton's day it was always a problem during hot summers and periods of drought to keep gardens properly watered. The use of rubber hose, which makes general irrigation at the present day so easy, did not start until the middle of the nineteenth century.(2)

It is always an advantage to have a stream or river running through a large estate, and it will be remembered that in 1730 Knowlton moved the kitchen garden at Londesborough, which was situated to the north of the house, to a new site in the valley where a mill stream could be diverted to run through the centre of the garden.(3) As a result he probably had more trouble with frost than on the higher ground, but how much easier it must have been to keep four acres of fruit and vegetables properly irrigated.

During the seventeenth and eighteenth centuries water from a spring, or river, situated near a garden, was conveyed into the property by means of pipes, whether lead, iron, earth, or wood. Wood pipes were made of Alder, Elm, Oak, and Beech. Those made of Alder were the cheapest though not the strongest. In order to collect a supply of water, many eighteenth century gardens has basins, or reservoirs, in the lowest part of the gardens into which water ran from the adjacent grounds in heavy rains. Besides pipes, water was also conveyed to places where it was required in large tubs or barrows, and the work of watering seeds, plants, shrubs, and trees, was performed in various ways, including watering-pots, watering-cans and pumps.

In a postscript Knowlton remarked:

'I Like Jaseck very well I beleve shall send him or my man to a place this autome

his Drying specimens has gone a grat way with ye Garden & will by S^t michelmus have a fine collection... because I cutt freely... without groudging [grudging] or sparing &c'

(1) Information supplied to the present writer by Richard Mortimer, of the Climatic Research Unit, University of East Anglia, in *epist.* 19.vi.80.

(2) Rubber hose was invented about 1827. A copy of Charles Macintosh's shop sign (1847?) lists garden and brewery hose, among the rubber products sold, and by about 1881 hose for watering gardens was among the commonalty of rubber products. (Information supplied by the Manager of the Dunlop Archive project in *epist.* 10.xii.79).

(3) *See* above, pp. 79–81.

It appears from the above passage that Jaseck was making a *hortus siccus*. Preserved in the Department of Botany in the British Museum (Natural History) are four sheets of herbarium specimens entitled: 'Plants from the north of England by Ld. Burlingtons gardener.'[1] Although not named this was in fact Thomas Knowlton. The sheets contain 15 specimens of Alder, *Alnus glutinosa* (L.) Gaertner (one of which is labelled in Knowlton's hand as being the Scarlet Alder), one specimen of *Arbutus andrachne* L., Andrachne,[2] two specimens of seaweeds, namely *Himanthalia elongata* (L.) Gray, and *Palmaria palmata* (L.) O. Kuntze, and two specimens each of two zoophytes which Knowlton would have thought of as seaweeds at that time. One of the zoophytes has attached to it the egg-case of a dogfish. These four sheets may have originally belonged to the *hortus siccus* referred to by the Yorkshire botanist James Bolton, in his *Filices britannicae; an history of the British proper ferns* [1785–] 1790. In writing of 'Alpine acrostichum', now named *Woodsia alpina* (Bolton) Gray, commonly called Alpine Woodsia, Bolton recorded:

> 'The specimen figured on plate 42, was brought from Scotland, but the plant is also a native of South-Britain; for in a volume of dried plants, collected by the late Mr. T. Knowlton, I have seen specimens of the same plant, with this note, in his own hand writing, "From the mountains of Wales".'

In 1887, the Director of the Royal botanic gardens, Kew, transferred the British herbarium of Thomas Knowlton to the Department of Botany, British Museum (Natural History).[3] This was probably lost in the fire during World War II. It may have been the volume of dried plants seen by James Bolton.

(1) *See* H.S. 240 fol. 113–116. J. E. Dandy credits folio 113 only, to Knowlton. *See Sloane herbarium,* under Boyle (Richard), 95 (1958).

(2) James Sherard is recorded to have been the first to cultivate this plant in Britain, *See* Aiton (W.), *Hortus kewensis*, etc., **2**, 72 (1789) under *Arbutus andrachne* L.

(3) *See The history of the collections contained in the Natural History Departments of the British Museum,* **1**, 114 (1904).

Arbutus andrachne L., *native of south-east Europe and the eastern Mediterranean region, was introduced into Britain from Smyrna in 1724. A specimen of it survives from what is assumed to be Knowlton's herbarium.*

TO SAMUEL BREWER

August [1737][1]
'S^r

I Recd y^r Baskett with ye plants as within mention'd all safe & I hope will do well; they apeare so now, wch I kindly thank you for...if I did not send you ye asarina [*Asarina procumbens* Miller, native of South France and North Spain] I have a plant att y^r sarvice when ever you please with any thing else I can spare'

In the next passage Knowlton wrote about an important publication by Leonard Plukenet (1642–1706), his *Phytographia* (1691–96), *Almagestum botanicum* (1696), *Almagesti botanici mantissa* (1700) and *Amaltheum botanicum* (1705), four books which may be considered as one great catalogue of plants with figures of new and rare species. The work was widely cited by contemporary and later writers, including Linnaeus and Sir J. E. Smith. The *Phytographia* itself appeared in four parts consisting of plates only and it may be considered to accompany the *Almagestum* which contains the specific definitions of several thousand species of plants, arranged alphabetically according to the generic names used by former authors. At the base of each plate in Parts 1–3 of the *Phytographia* are the names and synonyms of the plants depicted. The fourth part wants this valuable information. However, the figures are cited, as are those of Parts 1–3, in the *Almagestum*. The *Almagesti botanici mantissa* contains numerous observations and corrections relating to the *Almagestum*. The *Amaltheum* is a supplement to the earlier volumes. Some owners of the work added in manuscript at the base of the plates in Part 4 of the *Phytographia* the names and synonyms of the plants depicted, and also added in the margins of the pages of the *Almagestum* information given in the *Mantissa*. By so doing this important, but complicated, work was made much easier and quicker to use. We thus find Knowlton writing a letter in which he made the following request to Brewer:

'You was once so good to tell me y^t you would gett some Body to copey over Dr Plucknetts names or nots of Consell Sherrards if youl oblidge me...I shall be very willing to pay ye transriber because it may be much more correct then myne I Know it is a work of some trouble so would not desire it of you...'

(1) Knowlton corres. British Museum (Natural History).

A copy of the work in the Botany Department of the University of Oxford (Press mark Sherard 569) has the names of the plants on plates CCLI–CCCXXVIII (Part 4 of the *Phytographia*) and on plates CCCXXIX–CCCL (of the *Mantissa*) supplied in manuscript at the foot of each plate. This copy has a printed label inside the front cover 'E. Bibl. Sherard' and the bookplate 'E. Bibl. Botan. Oxon.', so the assumption is that the volume is from Sherard's library. Other copies of Plukenet's work with manuscript notes are the copies that originally belonged to J. J. Dillenius (Sherard 736) and to Leonard Plukenet respectively. Dillenius's copy is now preserved in the Botany Department, Oxford, and Plukenet's is in the Department of Botany of the British Museum (Natural History).

Knowlton, always showing himself willing to let people have any seeds or plants he could spare, was especially generous to Brewer:

> 'I am . . . well pleased you mak a Little cash of my plants for thirs no reson to give unless a sutable return w^{ch}. . . I hope shall allways be my . . . Inclynation; for when ever I desire or Receve it is generally such plants as gives me a pleasure & to all those I ever corresponded with I have endeverd to be as gratefull & generious . . . as I could . . . I do asure you I am so to all . . . I have ye Lesst correspondance with'

The writer continued by telling Brewer:

> 'when ever Smith comes over hither pray by a Line Lett me Know what it is youl want if it is ye tagetes [Marigold, African or French] Il send you a plant or 2 of it . . . if you think to send ye cerasuss [Cerasus](1) this season it will be much bett^{r} without a pott . . . I never Knew any plants in my Life come so well that way by carrers [carriers] . . . but if it be not prity well rooted it had much bett^{r} stay till spring My son [Charles] att Last is gone to Cockswould scholle neare to ye L^{d} faulking Bridges house &c about 30 miles from hence where I hope hel have all ye good Instructions nessary for a foundation to . . . passe thro ye seen or stage of Life agreeable thereto; I shall do what I can by jentle means to Engaige him a Love for nature in all I am able & have not other thoughts but to mak him a gardener if I can have ye lest Incoragement thereto'

The old Coxwold school is an attractive building still standing to this

(1) Cerasus: Cherry-tree, *See* Miller (P.), *The gardeners dictionary,* (1731).

day in the village street. 'L^d faulking Bridges house' was Newburgh Priory – which lies just beyond the east end of the village, and when the above letter was written it was owned by the Earl Fauconberg (d. 1774).[1] The school was a Free Grammar School founded in 1603 by Sir John Harte, Knight, Citizen and Alderman of London. A schoolmaster was paid a yearly sum for teaching the children English.[2] By 1818, but perhaps as early as Charles Knowlton's time, Latin and Greek were also regularly taught.

Knowlton told his friend about a parcel of seeds he had received from Italy which had given him little pleasure:

> 'I has a colletion of seeds by a gentleman from Itally they are good things but I have of y^m all but ye white amaranthoides & that I feare is ye worst seed in all ye whole parsell &c'

In the third edition (1737) of Miller's *Dictionary* the white and purple *Amaranthoides* were considered two different species. Later, however, they were shown to be one species only and this was named by Linnaeus *Gomphrena globosa,* Globe Amaranth. Although the seed was sent to Knowlton from Italy this plant is a native of tropical America.[3] It was first introduced into Britain from the East Indies in 1714.[4]

Finally Brewer is told:

> 'Mr Blackburn wrote me word his plantain [*Musa* cv.] has pushed a Large flower stem wch he is in grat hopes will perform ye operation of nature in Bring forth flower & frute...'

In a postscript, dated 'August ye 21', the gardener added: '... I am in grat hopes ye Dictamnus creticus [*Origanum dictamnus* L., Dittany of Crete] will do and Look well'

(1) Information supplied by the County Librarian, Northallerton, North Yorkshire, in *epist.* 16.v.78.

(2) Carlisle (N.), *A concise description of the endowed grammar Schools. Lancashire and Yorkshire,* 795 (1974) reprinted from the 1818 edition.

(3) Backer (C. A.) and Bakhuizen van den Brink (R. C.) *Flora of Java*, **1**, 239 (1963) Groningen.

(4) *Curtis's Botanical Magazine,* 2815 (1828).

TO SAMUEL BREWER

Octr ye 9 1737[1]
'S^{r}
I Recd all y^{r} papers wch I thank you for ye Solanum antiquorum is ye alkengij in authors. Tho I have not Looked in any Book for it but I Know it is ye same you take it for because... there is no other so calld & is a[n] old plant.'

The Solanum of the ancients to which Knowlton was referring is presumably Alkekengi, hence possibly *Physalis alkekengi* L., Winter Cherry. Knowlton also told his friend: 'Y^{r} flos passionis you call ye watt^{r} Lemon is not it neither Like unto it but pluck: Tab 211 fig. 3'. The authority here cited, namely Plukenet's *Phytographia,* Part 3, tab. 211, fig. 3 (1692), illustrates 'Passiflora arborea laurinis foliis americana', called by Plumier in 1693 'Clematis indica fructu citriformi foliis oblongis' and by Linnaeus in 1753 *Passiflora laurifolia.* This species is native of the West Indies and tropical South America. It is sometimes called by its common name of Water Lemon.[2] Although Plukenet included Plumier's name for this Passion-flower in his *Almagestum* of 1696, he did not give it as a synonym in the *Phytographia.* Furthermore, none of the other synonyms for this species in the *Phytographia* indicate the lemon-shaped leaf. In view of the inadequacy of Plukenet's figure in comparison for example with the illustration in Charles Plumier's *Description des plantes de l'Amérique* of 1693, and the consequent difficulty of being able to say whether a particular plant is or is not identical with it, the identity of Brewer's plant must remain in doubt.

It will be remembered that in an earlier letter Knowlton had told Brewer that if he was thinking of sending him a plant of 'cerasus' by the carrier, he would prefer it to come without a pot.[3] In the present letter he said:

'I have not Recd ye cerasus yett but w[ould] much rather have a plant without ye pott & am sa[tis]fied will come much safer y^{n} with wch I am convinced of by... ye Grate n^{r} Recd from Divers parts & sent my self without... & always shall'

Knowlton, being against sending plants in pots by the carriers for fear of

(1) Knowlton corres. British Museum (Natural History).
(2) Gooding (E. G. B.) and others, *Flora of Barbados,* 298 (1965).
(3) *See* above, p. 158.

the pots being broken and the plants damaged, preferred them to be sent with a little earth round the roots and packed in moss in a box.[1] There is scarcely any information available concerning the common flower pot of Knowlton's day although the material used in making such containers was almost certainly earthenware, which was used during the nineteenth century and, indeed, until the present day, when plastic became available.[2] It is likely, however, that the making of earthenware flower pots in the eighteenth century was more dependent on local clay supplies than those made at a later period and that the colour would therefore have varied, from yellow to shades of red, from one region to another.

Regarding the shape of early flower pots the French agriculturalist Louis Liger (1658–1717) directed that pots for all sorts of flowers should be 'three inches narrower at the bottom than at the top, to the end flowers may be the easier removed out of them.'[3] In 1824 Loudon wrote of the common flower pot as 'a cylindrical tapering vessel of burnt clay' sold in ten different sizes.[4] The shape, therefore, remained constant. A figure of a seventeenth century flower pot of such a shape was drawn by John Evelyn (1620–1706), author of important books on practical planting and gardening, including the celebrated *Sylva*. He made the drawing of a flower pot, together with many other sketches of garden tools, for his 'Elysium britannicum',[5] a work begun about 1659 but never completed.

It is interesting to note that in a letter of 17 November 1729 Knowlton refused an offer of some plants from Brewer for his new stove, because until he went to London and ordered some pots he had nothing suitable to put them in.[6] This seems to indicate that the gardener could not obtain flower pots locally. Indeed, David Neave, who has done a lot of detailed work on the Londesborough area, has pointed out that although at that time there were one or two brick and tile works in the region, it is unlikely that there would have been kilns making flower pots. It is possible that when he went to London, Knowlton purchased them from one of the seedsmen and nurserymen situated in and around the metropolis.

The difficulty in Knowlton's day of being able to water gardens

(1) *See* below, p. 165.

(2) For details of nineteenth-century pots *See* Loudon (J. C.), *An encyclopaedia of gardening*, etc., 284 (1824), and later editions.

(3) *See* London (G.) and Wise (H.), *The retir'd gardener in six parts... The four last parts... translated from the French of... Louis Liger... The second edition... by Joseph Carpenter.*, Part III, 148 (1717).

(4) Loudon (J. C.), *An encyclopaedia of gardening*, etc., 584 (1824).

(5) *See* Huxley (A.), *An illustrated history of gardening*, 116 (1978).

(6) *See* above, p. 119.

adequately during periods of drought has been referred to earlier.[1] Such times must have been very worrying to gardeners, and in the present letter Knowlton complained:

'... ye strawmonims this sumer have flowerd ye worst I have ever had y^{m} ye season being so continal Dry & scorching that very Little in y^{e} garding-way would grow...'

It appears that Brewer thought of including Plukenet's names of plants in the botanical guide he was compiling, and this led Knowlton to observe:

'... if you printe Pluck: names I cannot but say it will be ussfull to Lasey people & mak y^{r} Book... sell y^{e} $bett^{r}$...'

The gardener continued by writing to his friend on a variety of matters:

'... y^{e} nasturtium trifolium [*Cardamine trifolia* L., Three-leaved Cuckoo Flower, native of Central Europe] is very poore Likewise being eaten by y^{e} black flys so much y^{e} Tulips have been planted again some time ago... y^{e} [annual] popey I hope will do well... I had but one Cutting of ye uva marina [*Ephedra distachya* L., native of southern Europe and North Asia] ... phytolaca minor [*Rivina humilis* L., Bloodberry, native of the West Indies and tropical and subtropical continental America] should be in a glasse stove with a Little warmth — Mr Blackburn Horn'd accaia [*Acacia cornigera* (L.) Willd., a tropical plant known as Bull-horn] is diferant from myne his spines being shorter & thicker y^{n} myne; did you observe it — of ye amber tree I have a [pla]nt for you I mean ambarm spirai [i.e. 'Frutex africanus ambram spirans, or Amber-tree', later named by Linnaeus *Anthospermum aethiopicum*, native of South Africa]; ye Laurel Leaved thymaloides [*Pedilanthus tithymaloides* (L.) Poiteau, native of South America and the West Indies] is ye coccineo of severall authors but it Reachedly [wretchedly] done — I am very Glad to heare you have come to a hapy agreement with Mr Black[burn]: it will ad some what to y^{r} collection y^{e} plantain [*Musa* cv.] I am very glad to heare is in such perfection I thank you for y^{r} Description of it &c y^{r} tithymaloides [*Pedilanthus tithymaloides*] y^{t} died perhaps was not for y^{e} want of $watt^{r}$... but for want of warmth wch is y^{e} reson of many a dath in tender plants

(1) *See* above, p. 154.

Manchester as you observe is a very fine town I am as fond of it as you can be & for ought y[t] I Know when I Leave... Bussness I may go & Live thir... tho I tould you asphodellus [the species is not mentioned] was dead yett is now grows very well & will mak a fine plant my Statice is Likwis struck... y[e] rains & coole weather has so recove[red] y[m] y[t] no one would belive it. I have 2 plants of Euphorbium Antiquorum [*Euphorbia antiquorum* L., native of India]... is all that offers S[r] y[r] most obydant Humble Sarv[t] *Tho Knowlton.*'

After signing this letter the writer added a list of plant names, with the following request:

'These are y[e] cutt: [cuttings?] you drew for Isack if you should hapen to have any of y[m] & when you can propigat so as to part with any [it] would be agreable to me... but pray do not straiten y[r] self...'

The list consists of eighteen different species, including:

'Adatoda salicis folio [possibly the shrub named 'Adhatoda indica folio saligno' included by Herman Boerhaave in his *Index plantarum* of 1720, Pt. 1, p. 239. It appears under the same name, with the added sentence of 'the willow-leav'd Malabar-nut, with a white flower, commonly called the Snap-tree,' in Philip Miller's *Figures of... plants* **1**, 9, pl. 13 (1760). This native of the Canary Islands is now known as *Justicia hyssopifolia* L.]...
Anagallis monnella [*Anagallis monelli* L., native of south-west Europe]...
Springe cyclamon [*Cyclamen vernale* Miller, now considered the same as *Cyclamen persicum* Miller, native of the Aegean region]...
Phytolaca minor [*Rivina humilis* L., mentioned earlier in this letter]
alaternoides telephij folio [*Clutia alaternoides* L., native of South Africa]...
Euonymus *Æ*theopica [*Maytenus lucida* (L.) Loes., native of South Africa].'(1)

Knowlton is his turn promised his friend:

'Il send you severall plants you mention towards Spring for being small will be to soon to part with y[m] till y[t] time so what ever I have y[n] to part with you will be welcome... but if you think you can tak a jorney so far in y[e] Spring you may chuse for y[r] self...'

(1) This plant was first introduced into this country in 1722 by James Sherard, *See Aiton (W.), Hortus kewensis,* etc., **1**, 273 (1789), under *Celastrus lucidus* L.

Cyclamen persicum *Miller, Knowlton's 'Springe cyclamon' is a native of the eastern Mediterranean region. From this species most of the familiar greenhouse and indoor forms of* Cyclamen *have been derived.*

At the end of his letter Knowlton returned again to the subject of the damage that could be done to plants if sent in pots:

'Isack tell me you have thought to send some plants to winter heare if you do pray do not pack y^{m} up with potts wch will be in Danger of Bracking & be ye means to Destroy y^{m} ye onely way... would be with Little Earth to ye root packed in mosse in a Box... & youl not feare any harm if in... good condition — I am very Glad to heare you are so well as you have been it is a grate Pleasure to me so that I hope youl Labiour att ye Guide wch I shall Long to heare is allmost Ready for ye Presse I hartely wish youl all ye health & Happeyness that this world will aford to our fellow creature[s]... y^{r} most obydant Humble sarvant
Tho Knowlton
P.S.... I should be glad to Know where that Mr Nevile Lives y^{t} has Bult ye green house I have a Large n^{r} of orange trees to dispose of [at?] Beaverly if it happens you can Recom̄end y^{m}...'

It is likely that 'Mr Nevile' was a member of the family of that name who lived at Kilnwick, between Beverley and Driffield, from 1650, but it appears that the Neviles left Kilnwick by 1722,[(1)] and a study of Joseph Foster's *Pedigrees of the county families of Yorkshire,* seems to suggest that by 1737 the Neviles of Kilnwick would be living at Chevet Hall near Wakefield in West Riding.

TO SAMUEL BREWER

Octr ye 26 1737[(2)]
'Deare S^{r}
I Recd y^{r} Box of plants all safe & hope to furnish you with y^{m} in Spring in very good health with a Large n^{r} of other sorts when I shall think y^{m} safe to remove then what ever I have to spare shall post to you with ye Gratest pleasure... ye Box come to me without any derictions I supose [they were]... Brock or Lost from ye Box & y^{r} Lett^{r} w^{ch} [you] had send so long before I mett with att Beaverly post yesterday I am surpris'd att ye many Blunders that is maid by ye post masters'

It seems likely that the intended destination of the letter had been Market Weighton, which is about ten miles nearer to Londesborough than

(1) Ingram (M. E.), *Leaves from a family tree,* 10–11 (1951).
(2) Knowlton corres. British Museum (Natural History).

Beverley. Brewer may have sent this letter to catch the newly established postal service that went from York to Hull by way of Pocklington and Weighton and also presumably Beverley.[1]

Knowlton told his friend:

'ye Lett^{r} I Recd from ye jun^{r}. d^{r}[2] was about some seeds wch I had promised to send for to portingall [Portugal] wch you may Lett him Know are safely arrived & shall be sent... with ye exspunses as soon as I have y^{m} from Hull the merch^{t}. who sent for y^{m} I Recd a Lett^{r} from yesterday that they was come & to Know where to send y^{m} to they are ye Laurus regus Bearrys...'

The plant Knowlton called 'Laurus regus' was *Laurus indica* L., now called *Persea indica* (L.) Sprengel, Indian Bay. Philip Miller wrote of this tree in his *Gardeners dictionary* (1768), that it:

'...grows naturally at Madeira and the Canary Islands, from whence it was formerly brought to Portugal, where it has been propagated in so great plenty, as to appear now as if it was a native of that country... It is called by some the Royal Bay, and by others the Portugal Bay.'

It will be remembered that Sir Charles Hotham was the owner of a property at South Dalton, a few miles north west of Beverley. He also owned a house in Beverley.[3] Knowlton continued his letter by writing on a number of subjects, including a reference to Hotham's gardener:

'I see m^{r} [John] Scot to day Sr Charels Hothoms gard^{r}[4] who said y^{t} you promised to send him some ariculas [*Primula auricula* L.] & desired me to remind you of it & if you do he desires youl send ye names of each sort... directed to him att Beaverly by wch means hel have y^{m} safe – I am very glad to heare you are so well as you seem to bee I hope youl neglect not any oppertunity of forwarding ye Guide wch I so much Longe to see;

(1) Regarding this newly established postal service *See* letter, dated Dijon 27 February 1739, addressed by Sir Marmaduke Constable to his Chaplain, in *Constable of Everingham estate correspondence,* 1726–43 [Edited by P. Roebuck], The Yorks. Arch. Soc., **136**, 109 (1976).

(2) The name of the junior doctor is not given.

(3) *See* above, p. 104.

(4) 'M^{r}. John Scot, gardener to Sir Charles Hotham'. *See* List of Subscribers in Miller's *Gardeners dictionary* (1731).

not any man I think can performe it bett^{r} y^{n} y^{r} self...if you continue in health and able to undertak such anther jorney I shall be glad to see you att Londesburght in ye Springe y^{t} you may chuse for y^{r} self but...if you think it to grat a fateague you shall have sent you what ever I have too [two] of that youl name...ye Euphorbium antiquorum [*Euphorbia antiquorum* L., native of India] I had 2 plants in ye Spring when you was heare but so small perhaps you took no notice of y^{m} but are fine ones now...y^{r} cutting of mellianthus [possibly *Melianthus major* L., Honey Flower, or *M. minor* L., both native of South Africa] Looks well but feare it will not strick [strike] this season — I have some thoughts in feb^{y} to Buld a small stove one [on] purpos for ye pine aples some what Like unto D^{r} wights of manchester...if I continue in ye mind am now in my head has been much taken up that way — I do asure you upon my word that I never promisd to send mahogony [*Swietenia mahogoni* (L.) Jacq., Mahogany Tree] to eather ye D^{r} or his sarvant because I had Intended y^{m} for you...so could do no such thing haveing then not to spare to any besides y^{r} self so it must be a mistake...Indeed ye gard^{r} asked me more y^{n} once for it & ye Hermania; (probably *Hermannia grossularifolia* L.][1] but ye very grate obligation you have so offten layd me under duty & gratitude both...maid me com̄unite [communicate] &c what I could part with to y^{r} self...'

The above passage concerning the Mahogany Tree indicates the competition between gardeners to acquire new introductions. According to a manuscript given to William Townsend Aiton, Royal Gardener at Kew, by Thomas Knowlton junior, grandson of the subject of this biography, *Swietenia mahogoni* was introduced into this country 'before 1734, by Mr. Mark Catesby'.[2] As Thomas Knowlton the elder and Mark Catesby were friends, it seems possible that the latter gave the gardener seeds of this tree after his return from the Bahamas in 1726.

Knowlton ended his letter by telling his friend:

'I lately [received] some new seeds from Italy...ye Alca [Alcea] vivi crispa folio [possibly 'Malva foliis crispis' of Caspar Bauhin's *Pinax* (1623), now *Malva verticillata* var. *crispa* L., native of East Asia] & phytolaca minore [*Rivina humilis* L., Bloodberry, native of the West Indies, tropical and subtropical continental America] are grate raritys I hope to propigate y^{m}...my plants all Looks in very good health...Il save you a few seeds of ye Laurus rigus [*Persea indica*] that you may be maid hapey in ye same faver pray ex-

(1) *See* above, p. 151.
(2) *See* Aiton (W.), *Hortus kewensis... The second edition enlarged by W. T. Aiton,* **3**, 38 (1811).

cuse my Imperfect scrible haveing to [two] or 3 Long Lett^{r} one [on] the same subject to some frainds neare London conclude me S^{r} y^{r} most obydant Humble savant
Thos Knowlton'

TO SAMUEL BREWER

Feby ye 7 1737/8[(1)]
'Deare S^{r}
I Recd yrs & am glad to heare of y^{r} health... as for ye Duk de van toll & cloramon tulips as Co̅mon names I have had y^{m} in flower amoungs gard^{s} & ye flo[rists] as well Known by those names as any thing in Life;'

Duke van Toll, or Winter Duke as it was sometimes called, and Claremont, were early flowering tulips, and are found listed in John Rea's *Flora* (1665), Philip Miller's *Gardeners dictionary* (1731), and other works. Knowlton continued:

'I am sorrow to heare we shall Losse you out of yorkshire tho manchester is a fine place & no dought Plants would sell well thire but y^{n} consider ye grat charge to remove thither would hadly answer in severall yeares... I hope youl still continue y^{r} old Habitation till such time you have finished y^{r} Guid wch I Long so much to see; the china aster [*Callistephus chinensis* (L.) Nees] tho 2 years old is good... my new seed was not good this year [and] is ye reson I did not send you of it'

The China Aster was first introduced into England by Philip Miller who, in 1731, raised plants from seed in the Chelsea Physic Garden, and Dillenius, who had received seeds from Adrian van Royen (1704–69), the Professor of Botany at Leyden, gave a fine representation of the plant on plate 34 of the *Hortus elthamensis*.

Part 8 of Mark Catesby's *Natural history of Carolina, Florida and the Bahama Islands* was reviewed in the *Philosophical Transactions* for April, May, June 1736, **39**, 441. Knowlton, however, did not receive his copy of this Part until shortly before he wrote the present letter, when he told Brewer:

'I have Recd Mr Cateby 8 part wch with a cargo of some seeds just arrived

(1) Knowlton corres. British Museum (Natural History).

from Caralina being some of y^m^ from Mr Collison [Peter Collinson] thirs Tulip trees ... one [of] y^m^ being ye Laurell Leaved ye other Comon with some fine Sassafras Burreys & other things wch I designe you a part ...'

The three plants mentioned were all natives of North America namely, *Magnolia virginiana* L. (syn. *Magnolia glauca* L.), Laurel-leaved Tulip-tree,[1] now commonly called Swamp Bay and Sweet Bay; *Liriodendron tulipifera* L., Common Tulip Tree;[2] and the berries of *Sassafras albidum* (Nutt.) Nees, White-flowered Sassafras. When Knowlton wrote of 'm^r^ Collison' he was referring to Peter Collinson who was instrumental in introducing so many new plants into England, especially from his American friend John Bartram who lived in Pennsylvania, by the Schuylkill river, near Philadelphia.

On the subject of Catesby's publication Knowlton also wrote:

'Dr Richardson has att Last I supose Bought m^r^ Catesby 8 parts of his nat: Hist: of Coralina ... from ... a Bookseller ... if ye D^r^ had a mind he had bett^r^ had it of ye author who would some times oblidge him wth some forain seeds as they come to hand.'

About the time the gardener wrote this letter to Brewer, the books, manuscripts and specimens belonging to the eminent Welshman Edward Lhuyd (also spelt Lloyd or Llwyd) were being disposed of, and we find Knowlton writing to his friend:

'... if I had time I have now ye Liberty of seeing ye whole collection of m^r^ Lhoyd 1300 manuscripts specimens & what ever he had collected togather ... D^r^ faulk ... a weltch man ... had all these I now mention under his care by Dyd 8n or 10 years ago in his native country & I have Latly had Leave to come & stay a month to Look over y^m^ ... ye printed Book excepted wch are just now valled & agoing to be sould to a seartin duk & will amount to between 5 & 6000 but the other is not regarded rather look one as Rubish tho to me ye most beautifull I feare I shall not be able tho so grat a temptation ye Duk of Bedford is ye person has ye refusall [they are] ... being valled by m^r^ quil [or qail or gail] & Davise & m^r^ Curt a gentleman y^t^ is writing ye antquites of England ...'

(1) 'Commonly called Small Magnolia, or Laurel-leaved Tulip-tree' *See* Miller's *Gardeners dictionary ... The eighth edition,* (1768).

(2) 'Common Tulip Tree', *See* Aiton (W.), *Hortus kewensis,* **2**, 250 (1789).

In view of Knowlton's dispostion to phonetic spelling the identification of the three valuers will always be suspect unless verified by manuscript sources. 'm^r Curt' is almost certainly the historian Thomas Carte (1686–1754) who used Lhuyd's manuscripts for his four volume *General history of England* (1747–55) and who would have been well qualified to be a valuer of Lhuyd's books. The booksellers, however, are more difficult territory. Mr Gwyn Walters, Assistant Keeper in the Department of Printed Books in the National Library of Wales, who has made a study of the disposal and dispersal of Lhuyd's collections, has suggested to the present writer,(1) that if he had to make a guess regarding the identity of 'Davise' his choice would go to the London bookseller Charles Davis, whom Timperley referred to as 'one of the earliest booksellers who retailed libraries by marked catalogues',(2) or possibly to Lockyer Davis, who was enormously respected in the trade and by the literary brotherhood. Regarding 'm^r guil [or qail]' there is no known bookseller in that period whose name begins with Q. The only real name it reasonably resembles is of Edmund Curll. But as pointed out by Mr Walters 'Curll seems an unlikely bedfellow for Davis, being a bad lot and always in trouble for pirating and other trade indiscretions. Moreover, Thomas Carte is known to have invoked a House of Lords Order or Injunction against Curll's practices in his own fight against Irish pirates. So, despite the name, Curll is a highly unlikely choice.'

Pulteney in his *Historical and biographical sketches of the progress of botany in England* **2**, 113–14 (1790), writes of the sale referred to by Knowlton, as follows:

'From private information, I have learned that Mr. Llhwyd left a very considerable library, a large collection of manuscripts and specimens; that in the year 1728, these were all in the custody of Dr. Fowlkes, of Lhanher, who died soon after that time. They were undisposed of ten years afterwards; but the printed books, which were of great worth, had been valued by a gentleman and two booksellers, and the refusal of them offered to the Duke of Bedford. My intelligence reached no farther.'

'Dr. Fowlkes, of Lhanher' ('Dr. faulk' as spelt by Knowlton), has been identified as Robert Foulkes, M.A., Rector of Llanbedr, near Ruthin, who shared with Lhuyd an interest in botany.(3) He was one of the early

(1) In *epist.* 20.xii.79.
(2) Timperley (C. H.), *A dictionary of printers and printing,* etc., 695 (1839).
(3) Information supplied by Dr Brynley Roberts of the Welsh department at the University College of Wales, Aberystwyth, in *epist.* 20.iii.78.

subscribers to Lhuyd's *Archaeologia britannica*, and some of his correspondence with Lhuyd appears in R. T. Gunther's *Life and letters of Edward Lhuyd (Early Science in Oxford,* **14**. Oxford 1945).

Edward Lhuyd (1660–1709) eminent in botany, and in many branches of archaeology, was also outstanding as a Celtic philologist. From 1691 until his death in 1709 he held the appointment of Keeper of the Ashmolean Museum, Oxford. He travelled widely and made vast collections of plants, fossils and minerals. In 1699 his celebrated catalogue of British fossils was published, illustrated with engravings, and entitled *Lithophylacii britannici ichnographia.* In 1709 the first part of his *Archaeologia britannica* appeared. Lhuyd was also responsible for many papers published in the *Philosophical Transactions.* He formed a valuable and extensive collection of Welsh, Irish, and other manuscripts and a fine library of books and drawings.(1)

In 1760 a new edition of the *Lithophylacium* was published by the antiquary Reverend William Huddesford who, as Keeper of the Ashmolean Museum from 1755 until his death in 1772, had Lhuyd's fossils under his charge at the Ashmolean. It is interesting to note that in the preparation of the new edition of Lluyd's treatise, Huddesford wished to consult 'a collection of letters to Lluyd' in Knowlton's possession,(2) and it seems likely that the gardener purchased these papers at the above mentioned sale of the Welshman's books and manuscripts in 1738.

After giving Brewer the details concerning the above sale Knowlton continued:

'I have not seen m^{r} [John] Scot this 3 months his master is Dead so I supose thire be Little done now wch has accasiond a grat strugle betwen charles pelham Esqr & Sr Robt Hillard for a member & my man being a freeman & ye esq^{r} my fraind & an anticourtur have elected him with some dificulty...

John Scot had been gardener to Sir Charles Hotham of South Dalton,

(1) For further details concerning the life and work of this eminent Welshman *See* Emery (F. V.), *Edward Lhuyd F.R.S. 1660–1709 (1971),* Cardiff, and for a thesis showing that it was Lhuyd's father and his gardener, the skilful botanist, Edward Morgan, who first put him 'in touch with the Scientific method', *See* Roberts (B. F.), *Edward Lhuyd, the making of a scientist* (1980) Cardiff.

(2) *See* two letters, dated respectively 28 July and 24 November 1759, addressed by the Revd. W. Huddesford to E. M. da Costa, published in Nichols (J.), *Illustrations of the literary history of the eighteenth century,* etc., **4**, 468, 469 (1822).

Tory member of Parliament for the borough of Beverley. Hotham died in January 1737/38. Subsequently, on 20 February a by-election took place when two candidates contested the seat. They were Charles Pelham, a Tory, who had inherited property in Beverley from his uncle Sir Michael Warton, and Sir Robert Hildyard, Bart. On the day of the election Pelham received 432 votes and Sir Robert Hildyard 389.[1] Knowlton was evidently a supporter of the Tory party.[2]

In the present letter Knowlton answered a request from Brewer to exchange gardeners. He wrote as follows:

'...to oblige you I will exchange Josek Roberts with you for a month but Longer I cannot well he's my chefe man to ye stove I cannot think of trusting so much care to y^r^ man without he had ye Like exspearince...perhaps I run a very grat Haszed in ye mean time my plants are all in Good order... I would have you not to fail to send y^r^ man over to be heare...sunday or munday next & as soon as he comes Josek will mak his best way over to you & if you have a mind for any thing he can Bring you...you may have it pray excuse me for not answering y^r^ Querys in y^r^ Last because I had not time to Look into Books & site ye paiges for you can hardly Imagin how much my time has been taken up in on thing or other...'

In a postscript Knowlton added:

'...the Taplo [*Crateva tapia*] you had of me...I raisd a grat many one yeare & have seeds or Berrys by me now...they are deficulest plant to keep as I ever had so are all Dead ye description was coped from m^r^ Catsby ...with no mention of Sr Hans [Sloane] I evry day exspect a cargo of seeds from afraca & Barbara [Barbary?] they are with ye carrers from London'

Mark Catesby in his *Natural history of Carolina, Florida and the Bahama Islands,* **2**, 32 (1734), gives a description of 'Tapia arborea triphylla', a tree named by Linnaeus *Crateva tapia,* and commonly known as the Garlic Pear Tree on account of the strong smell of garlic possessed by all parts of the plant. It is a native of the West Indies and tropical America. The passage

(1) Bean (W. W.), *The Parliamentary representation of the six northern counties of Yorkshire,* etc., 743 (1890).

(2) Regarding Knowlton's politics see also below, p. 199.

in Knowlton's letter concerning this tree shows once again how much he studied books. He copied the description from Catesby's work but was quick to point out that this naturalist made no mention of the description of this tree in Sloane's *Voyage to . . . Madera, Barbados, Nieves, St Christophers and Jamaica,* **2**, 169–70 (1725).

TO SAMUEL BREWER

July ye 6 1738[1]
'D^{r} S^{r}

I Recd y^{r} oblidging Lett^{r} with the Inclosed I am sorrey it was so Long ye parsell before . . . came to y^{r} hands I hope ye sensative [plant] will grow [*Mimosa sensitiva* L., native of Tropical America]

as to Isack I do not remember that I ever asked ye faver of you to take him from you, it was your own offer & Desire to do hime some sarvice . . . no man in ye world I am shure would be so ready to sarve you as my selfe Therefore hope youl not think I took any Indrict means to tempt him away from you . . . I read y^{r} Lett^{r} to Isack he tells me he has all ye gratitude & Duty to you for y^{r} so good a Kindness & he says he shall bee willing to mak you all ye recompene & retaliation he can . . .

I have wrot to a young gentleman who spook to me about ye 3 of may of a young spark to come to me to Larn gardening when I have seen him & talkd will Lett you Know as you so much Desire it – for my part I never pertent [pretend] to help young mean to places but such as . . . I can faithfull[y] recom̄end & gett y^{m} good waiges & have thanks & creaditt by y^{m} – & besides can gett y^{m} much bett^{r} waiges than they can themselves – & as to you I feare it can not so farr fall in my power to sarve you – you Know if they work will I can afford y^{m} good waiges wch is a temptation –'

From this remark it seems that Brewer had asked his friend if he could get him a gardener. He continued:

'. . . ye Lds Gardener y^{m} mention was under the gratest obligation to me I gave him I belive 46 or 50 pine aples plants such as you had & since 24 more wch he sent over for & he in ye Gratest narsity [necessity] for y^{m} to furnish him an other yeare – Ld Bathhust [Allen Bathurst Lord Bathurst, later Earl Bathurst] had to [two] fine frut of me Last sunday & one [on] Tusday ye Duck & Duches of Cleaveland [William Fitzroy, third Duke of Cleveland and his wife Henrietta] had 4 more who are going to arraby castle

(1) Knowlton corres. British Museum (Natural History).

[Raby Castle, County Durham, the residence of the Duke's nephew the Earl of Darlington] with m^r^ bain & I shall cutt contualy n^rs^ of y^m^... perhaps as good in flaver... to thos from abroad – the D^r^ I Recd the plants from was not of y^r^ aquaintance but was Dr Chambers of Hull who I have sent 100 & od specimens of plants to &c...'

The gardener further informed his friend:

'...if you live & go one [on] with ye Garden I hope to have somewhat to please you next spring Difrant from what you have but as my plants are so small cannot yett Increase... y^m^... I am now in my gratest Hurry haveing bett^r^ y^n^ 40 people to govern & take care of; besides very frequant intruptions &c wch I hope will excuse me... writing... my sons att schoole & has been some monthes & hope hes well my wife joyns in complements to you who am S^r^ y^r^ most obeydant Humble sarv^t^
Tho Knowlton
y^r^ 2^d^ sort of ficeoides [*Mesembryanthemum*] I have Likewise but can hardly belive there is so many as Dr Dilinious [Dillenius] maks in ye H[ortus] E[lthamensis](1) I send you seeds of Tagetus so you may see ye way of flowering &c it is not in Millers Dict[ionary] it is in Hortus E[lthamensis] by the name of Tagitis flore multiflore minima [i.e. 'Tagetes multiflora, minuto flore albicante',(2) now called *Tagetes minuta* L., a native of South America] &c talboom or talbo is not in ye Hortus Elth but in H Am: [i.e. Jan Commelin's *Horti medici amstelodamensis rariorum plantarum,* Pars 1, 73, pl. 37 (1697)]... the Talbo is an old acequaintance of mynse & is a prity tender stove plant...'

The correct Sinhalese name is Tolabo, not Talbo as given by Knowlton. The modern Latin name is *Crinum zeylanicum* (L.) L.,(3) native of Ceylon.

Finally, Knowlton ends his letter as follows:

'I hope to have cutting strik of ye arbor Lavendulafolio [*Kleinia neriifolia* Haw., native of the Canary Islands] wch as soon as I have shall sen unto you with many other things if you will not come & fetch y^m^ y^r^ selfe... Isaca Says he is very willing to pay you... 20 shilling & if that will please you Il retain it in my hands till I have an opertunity to return it with my subscrib-

(1) There are more than fifty kinds of *Mesembryanthemum* illustrated in the *Hortus elthamensis, See* **2**, pls. 179–215 inclusive.
(2) *See Hortus elthamensis,* **2**, 374, pl. 280.
(3) *See Taxon,* **30**, 481–82 (1981).

sion money if you ordur my book to be bound in London will be bett^r the other may come in sheets July ye 7 by this days post I have a Lett^r from m^r Catesby who says that all ye greenhouse & stove plants [at Eltham] are disposed of gratis to thos who could make ye most Intrest for y^m to m^rs Sherard...'

This was the widow of James Sherard who had died a few months earlier on 12 February 1737/38. It appears that after Sherard's death the plants in the garden at Eltham were given away by his widow to anyone who wanted them.[1]

TO SAMUEL BREWER

August ye 13 past 4 of ye clock [1738][2]
'S^r

I had y^r faver by Smith[3] & have sent you 6 genoa narsus [narcissus?] & 4 others of ye Dutch trib[e] Likewise 1 Sugar Cane [*Saccharum officinarum* L., a species that 'originated on islands in the South Pacific, most probably in New Guinea'[4]] 2 sorts of flos passions [i.e. 2 species of the genus *Passiflora* L., commonly called Passion Flowers]... ye Catalonium & Itallian jasmines [i.e. *Jasminum grandiflorum* L., native of Arabia and *Jasminum humile* L., native of the Sino-Himalayan region]...'

The Botanical Register, **2**, 91 (1816) records that *Jasminum grandiflorum* acquired its common name Catalonian Jasmine from being:

'... cultivated in Catalonia, especially in the neighbourhood of Barcelona, to a greater extent than in any other part of Europe... It occurs likewise in almost every garden in Valentia, Murcia, and Andulasia, where, as in Catalonia, it is completely domesticated, thrives in the open ground the year through, and is hardly ever without bloom.' *The Botanical Register,* **55**, 350–51 (1819) records: '*Jasminum humile*... is frequently called, *Italian* yellow Jasmine by the gardeners, the plants being annually brought from Italy, by those who come over with Orange Trees.'

(1) *See* below, p. 197.
(2) Knowlton corres. British Museum (Natural History).
(3) Probably the nail man in Bradford, named Henry Smith who, at times, delivered letters and goods to, and from, Knowlton and Brewer, *See* above, p. 126.
(4) Purseglove (J. W.), *Tropical crops. Monocotyledons,* **1**, 215 (1972).

Among the other plants Knowlton sent to Brewer were:

'... a genistella montis [*Genista hispanica* L., Spanish Furze, native of northern Spain and southern France] &c Elychrism or: [i.e. Elichrysum orientale,(1) named by Linnaeus *Gnaphalium orientale* and now called *Helichrysum orientale* (L.) DC., commonly called Immortal Flower, native of Greece and the Aegean Region] you so much wanted Likewise a plant with flowers Like unto ye antirhinum...'

Finally, Knowlton told his friend:

'I have not yett Rec'd y^r^ plants what ever they are I had a lett^r^ from orford Hall [the residence of John Blackburn near Warrington, in Lancashire] wch tells me ye plantain [*Musa* cv.] is now pushing forth a flower the season now is so frefreshing that Londesburght Park is... as gay as is possaible.

The nail man [Henry Smith?] talkes... [that] m^r^ Rawson of Bradford wrot to me for some orange trees if he did I never Recd his Lett^r^ wch he might depend one [on] being answerd

I am in Grat hast y^r^ Real fraind & Humble sarvant
Thos Knowlton'

TO SAMUEL BREWER

Sept ye 13 [1738](2)
'D^r^ S^r^

I Rec'd y^rs^ but being in so very Grat trouble by my dearest betty & wifes Ilness & Death...She was take... of a violant feaver wch in a weeks time carred her of[f] to my very grat grefe & sorrow... as Iasek & you cannot determine [the sum to be paid] I have not sent any cash by Smith... till he & you determine [what to do]... Il medle no more twas y^r^ Intrest & fraindshipe recom̄ended him to me; hes not ye most oblidgeing person I ever had...

In this confusion I could not think of any thing to oblidge you; it was morraly Impossable for me so hope youl excuse it; yesterday m^r^ [Henry] Scot his Ld Ships [the Earl of Burlington] gardener of chiswick was hear

(1) This name is given by Caspar Bauhin in his *Pinax theatri botanici,* 264 (1623), Basileae Helvet.

(2) Knowlton corres. British Museum (Natural History).

&c m^{r} [John] Scot Sr Charles [Hotham's](1) Gard^{r} was att my beloveds feunerall... hes had a grat n^{r} of ariculars [*Primula auricula* L.] & polyanthus from ware [in Hertfordshire?] & other parts but he says he has Lost some of y^{rs} wch he seems to seat ye Gratest vally one [on] I am in ye gratest hast y^{r} Sincerly

Tho Knowlton'

TO SAMUEL BREWER

Tusday Septr ye 28 [1738](2)
'D^{r} S^{r}

I wrot a Longe Letter soon after smith [the nail man from Bradford?] was over heare & one by him with some plants I sent along with him & since I have not heared a word from you & Isack haveing aimed to see you & ye Rest of his frainds brings this to wait one [on] you – I heare by m^{r} Tilford juner [possibly John Telford (1689–1771) a member of the famous firm of nurserymen of that name at York] that you have been att manschester & I hope att orford hall [i.e. Blackburn's residence in Lancashire] if so I bege youl oblige me with a perticulare of all y^{r} jurnery – I had a Lett^{r} from worsope man^{r} [in Nottinghamshire, belonging to the Duke of Norfolk] this day that L^{d} petrea [Petre] was to be thir... ye same day... Mr Switzer [nurseryman and garden designer]... has been down to atend his Grase...'

Lord Petre was visiting Worksop Manor, the residence of Edward Howard, Duke of Norfolk (1686–1777), where he had prepared plans for an ambitious layout for the park and gardens. Some of his designs survive and are preserved in the Essex Record Office. They include a detailed planting scheme, signed and dated 24 March 1737/38.(3) Stephen Switzer (d. 1745), one of the most noted authors on gardening during the first half of the eighteenth century, was also a garden designer of considerable eminence, and he was possibly assisting with the major work being carried out at Worksop.

(1) i.e. the late Sir Charles Hotham, who died in January 1738, *See* above p. 104.
(2) Knowlton corres. British Museum (Natural History).
(3) *See* Binney (M.), Worksop Manor, Nottinghamshire, *Country Life,* **153**, 680–82 (1973).

'...by florist Isack [continued Knowlton] I have sent you ye orleana [*Bixa orellana* L., Annatto] you so much Desired...but then you know I had promised my old plant to Mr Blackburn...'

Concerning this handsome shrub we read in Curtis's *Botanical Magazine*, **35**, 1456 (1812), as follows:

'Native of South-America, the West and East-Indies; at least it is cultivated in these parts of the world, chiefly for the sake of the colour extracted from the seeds, and which is known here by the name of Anotta, or cheese-colouring, being the drug commonly used in Gloucestershire to give a deeper colour to their cheese. It is said too to be added to their butter, for the same purpose, by the Dutch dairy women. The Indians paint the whole of their bodies with this colour; and it is sometimes used as a rouge for European females. The South-American Spaniards make great use of it as an ingredient in their chocolate and soups, not only for the sake of its colour, but on account of its supposed cordial virtues.'

Knowlton closed his letter thus:

'I am in has[te] as I am offten S^{r} y^{r} most obydant Humble sarvant
Tho Knowlton
P.S. when ever you can spare y^{r} coppy of Mr Sherrards plucknett I would tak it as a faver...'

It will be remembered that the gardener wished to have a transcription of the manuscript notes in William Sherard's copy of Plukenet's work of 1691–1705.[1]

TO SAMUEL BREWER

Oct ye 3^{d} 1738[2]
'Deare S^{r}
I Recd yrs of ye 24 of Last month...I thank my God I am very well, considering ye grat Loss I have had in one of ye Best of wemen & wifes...
I am realy sorrey Smith should behave in so very Indecent maner to you...ye Lett^{r} I sent by him I did not Exespect he would ask any thing for

(1) *See* above, p. 157.
(2) Knowlton corres. British Museum (Natural History).

it because I allways take care to mak much of him & desred he would Leave it att ye posthouse till you sent thire for ye others . . . if you are resolved to Leave Bierly & steare towards manchester I shall tell you I will be sorrey for it; but if you are in want of Iosack & have a mind [to] go amoungest y^{r} tennants in Wilts[1] you shall have him when ever you please for 1 month or too but Desire youl give me as much notice as you can

I had a Lett^{r} from albin [Eleazer Albin] who teles me he has published a 3^{d} vollom of Birds [i.e. his *Natural history of birds*] as d^{r} stanope hase it should be glad to heare how you Lik it before I send for it . . .'

Although Knowlton's name does not appear as a subscriber to the first volume of Albin's work, it is found among the list of subscribers to the second and third volumes. Dr Stanhope lived at Ecclesfield, near Bradford and it would not have been difficult, therefore, for Brewer to go and inspect the doctor's copy.

Knowlton had apparently written to Lancaster for some plants. When they arrived, packed in two large boxes, they consisted of 'such Little Bits' he could do little with them. One of them was 'Digitalis acanthoides', now called *Isoplexis canariensis* (L.) Loudon. The gardener continued his letter by telling Brewer:

'. . . as to oranges & Lemens Il send you of each sort when ever you please & a stripe Lemon [possibly *Citrus limon* (L.) Burm. f., cultivar with variegated leaves] Likewise which is not in y^{r} naiborhood . . . asarina [*Asarina procumbens* Miller, native of south France and north-east Spain] I have plenty, so you may have it when every you please — his L^{d}Shipe being heare I have cutt a Large n^{r} of pineaples some of y^{m} very Large & fine & have still a considerable n^{r} to spare to sarve his taible he talks of staying till Xmas before he will seat forward for London

m^{r} baine att Raby Castly is Bulding a stove for pine apples & Likewise another att Chatsworth & one more in Lincolonshire where my man Jno is gone too [to] Live so that ye pine aples will gain very much in this Island in a few yeares time'

Knowlton also wrote to his friend, 'no dought you see in the newspapers ye Death of d^{r} Borhave [i.e. Hermann Boerhaave, Professor of Medicine and Botany at Leyden] . . .'

(1) Brewer had a small estate at Trowbridge, Wiltshire, *See*, above p. 92.

TO SAMUEL BREWER

9ber [November] 18 or 20th [1738][1]
'D^{r} S^{r}

I Recd y^{r} faver with m^{r} millers [possibly Philip Miller, Gardener to the Society of Apothecaries at the Chelsea Physic garden] 2 Rects wch I shall take care to send pay for att the first opertunity — I wonder you should offer anything for ye Lemons & oranges I tould you before y^{t} [I] Intended y^{m} as a presant unto you so desire youl except... y^{m} with out more cerimoney...as to Smith I bege youl not any more trouble me with him he is not worth y^{r} wile nor mine to spend time about[2]...I...am very sorrey to heare of any misfortuns in y^{r} famely it is what to offten happens to all people att one time or ye other...'

Details of Brewer's misfortune are not known. As already mentioned, he often had problems in his life; for example, in a letter dated 30 January 1727, J. J. Dillenius wrote to the eminent Swedish naturalist Carl Linnaeus:[3]

'I have had frequent letters from Mr. Brewer, and his last is so melancholy I am afraid the poor man will die... The return of his son from the Indies, who was dismissed the service, brought great trouble upon him, and he had been ill ever since.'

In a letter, dated 13 January 1728, and written by Brewer himself, to Sir Hans Sloane,[4] we read:

'...the discouragment I met wth from... my misfortunes is a great check to my inclinations — great loss by Fier bad debts some thousands by a partner[5] & being barred by an elder Br I & other relations from 3 or 4 considerable estates by unjustly cutting of entaile are shocking things to think upon.'

Some years later, on 20 August 1733, when Brewer was living at Bierley,

(1) Knowlton corres. British Museum (Natural History).
(2) Henry Smith, the nail man from Bradford? *See* above, p. 126.
(3) Smith (Sir J. E.), *A selection of the correspondence of Linnaeus,* etc., **2**, 144 (1821).
(4) Sl. MS. 4050 fol. 36. in British Library.
(5) Possibly a reference to his 'woollen manufactory' at Trowbridge, in Wiltshire, *See* above, p. 92.

he wrote to Sloane again, on the same subject, telling him:[1]

'I... dayly suffer very much by uncommon misfortunes wch could not be foreseen or prevented amounting to more than twenty thousand of pounds of my owne acquiring by the sweat of my browes, & to compleat my ruin my elder Bro^{r} has barred [me] from a considerable estate wch my Father settled upon me in consideration of some hardships he did me having entailed his estate.'

Of further worries Brewer recounted:

'I met wth some bitter losses since I was here [i.e. Bierley] wch pinches harder then thousands did before, & a late undutifulness of a daughtr has disordered my circumstances & given me more trouble y^{n} all my misfortunes beforementioned.'

Towards the end of his life Brewer had to contend with increasingly poor health and when the Reverend Richard Rauthmell who resided at Browsholme Hall, the home of the Parker family in West Riding, Yorkshire,[2] wrote to Dr Richard Richardson, on 2 September 1737, he told him:[3]

'Mr. Brewer... has extensive correspondence, still adding to his garden; but I thinke he grows weary of every thing else; for I very seldome see him.'

To return to Knowlton's letter to Brewer, the gardener told his friend:

'... our L^{d} talks not of going to London till about Xmis wch engaiges me very mucch so that I cannot answer Lett^{rs} now so readly as att... more Leasure

I very Latly have Recd very oblidging Lett^{r} from Mr Catesby who tells

(1) Sl. MS. 4053 fol. 30.

(2) *See Extracts from the literary and scientific correspondence of Richard Richardson*, etc. [edited by D. Turner], 315 footnote (1835). Browsholme Hall is three and a half miles north west of Clitheroe.

(3) *ibid.* 360.

me ye Last n^{r} of ye transactions[1] is worthy my reading so will send it me down with one of ye Large Lawerl Leaved Tulipt tree [*Magnolia grandiflora* L., Great Laurel-leaved Tulip Tree,[2] now commonly called American Bull Bay]...'

Knowlton also informed his friend:

'...he [Catesby] was Latly att L^{d} peatreas [i.e. Lord Petre] where is such a verity... and abundance... to Longe to Relate in a Latt^{r} — I have by ye same post a Lett^{r} from my L^{d} pettra gard^{r} [James] Gordon who is about Leaveing him & settling some where about Mile End in ye way & maner [the nurseryman] m^{r} fairchilds was att Hoxton he wrot to me if I could part with a few pine apples plants wch he intends to Lawnce out with, but I feare [the project] will not answer so well... but I wish it may he haveing married a fortune to enaible him therein — he writes me word that [he] had 2 or 3 sorts of papays [*Carica papaya*] in flower & that he had a palme just comeing to flower with a grat n^{r} of other fine things...'

James Gordon (1708?–80) became as famous a nurseryman as Thomas Fairchild. His garden was at Mile End in the parish of Bromley and he became outstanding for his skill in raising all sorts of foreign plants from cuttings, roots, and layers. Peter Collinson, the great promoter of botany and horticulture, wrote of Gordon in a letter addressed to Brewer, and dated 2 June 1742, as follows:[3]

'Our old Gardners are all gone off so that Curious & Rare plants are gott into private hands – Butt wee have one Lately raised up – that Is James Gordon a most knowing and Ingenious man who has had great Experience being Gardner to Mr Sherrard[4] & Ld Petre. He has taken up the Cudggels I do all I can to assist him Doc. Perkins can give you an account of his present situation who ever wants seeds or plants may be sure of being Honestly served.'

(1) The *Philosophical Transactions,* for the month of August and September 1738, No. 449, Entry XI. This number gives details of plate 61, entitled 'Magnolia altissima, flore ingenti candido. The laurel-tree of Carolina' (*Magnolia grandiflora.*).

(2) Named thus, in *A catalogue of the plants in the garden of John Blackburne, Esq. of Orford, Lancashire... By Adam Neal, gardener,* 41 (1779).

(3) Bound at end of Knowlton corres. in British Museum (Natural History).

(4) James Sherard of Eltham in Kent.

American Bull Bay (Magnolia grandiflora L.), *native of the southern United States, was introduced into Britain before 1732 and until about 1900 was the only truly evergreen hardy* Magnolia *in cultivation.*

Knowlton did not continue his letter until 'Xber [December] ye 8th', when he wrote:

'... ye above was written Longe before ye Recet of y^{rs} & I Designed to have sent you a Box of trees &c but as y^{r} unseartanty of staying is such I have forbore till I heare from you in Springe(1)... I shall be glad to heare how Dr Richardsons stoves & pine aples go one [on], I have abundance of fine frut still wch is a grat Pleasure to his L^{d}Shipe & creaditt to me... I am apt to belive my new pineaple fram[e] or Bead beats all that ever I see yett for that purpose & is much less Exspence then any thought of... I shall rejoyce to heare of y^{r} healthe & contnuation in yorkshire... my adatoda salsigna folio [possibly *Justicia hyssopifolia* L., a native shrub of the Canary Islands](2) by standing [in] ye green house I feare I have losst... I shall be extreemly sorrey for I realy did not Im̄agin it was so tender haveing allways thought it to be a green house plant... our Ld Continues heare still so that he talks nothing of going away yett wch has maide me so Long in answering y^{r} Lett^{rs}– besides I exspect Sr Jno Kay over heare to have waited one his L^{d}Shipe with ye young Dr Richardson...'

The first of these persons was Sir John Lister Kaye, Bart. (1697–1752) who, in 1730, married as his second wife Dr Richard Richardson's eldest daughter, Dorothy. He was Lord of the Manor of Thornton-in-Craven, about five miles south west of Skipton in the West Riding of Yorkshire.(3) Knowlton closed his letter with the following request:

'I should be glad to heare that Dr Stanope has Mr Albins Book of Birds i.e. ye 3^{d} voll; & how he & you Like it because as I tould you before I had a Lett^{r} from him some time ago about it...'(4)

It is not possible to be certain to whom Knowlton was referring when he wrote in the above letter of 'ye young Dr Richardson'. It may have been to Dr Richard Richardson's son, Richard (1708–81) or to his son named William (1710–83). Both men received the bachelor's degree in 1736 from the University of Cambridge. Richard was of Pembroke College, while

(1) Possibly a reference to Brewer's proposed move to Manchester. *See* above, p. 163.
(2) *See* also above, p. 163.
(3) *The history and antiquities of the deanery of Craven in the County of York... third edition... edited by A. W. Morant* (1878).
(4) *See* also above, p. 179.

William was of St John's College. Later, in 1741, William took his doctor's degree in medicine, but this was after Knowlton wrote his letter. Subsequently he settled at Ripon where he practised as a physician. He is said to have inherited 'his father's taste for botany and natural history'.[(1)] Previous to receiving the bachelor's degree, Richard had been admitted on 21 October 1730 as a student to the faculty of medicine at the University of Leyden, but there is no record that he received a doctorate.[(2)]

Equally ambiguous as the reference to 'ye young D^{r}. Richardson' is the form of address on the next letter to be quoted. This, dated 18 November 1739, was addressed by Knowlton:

'To
the worthy Dr Richardson
junior at north Bierly
Yorkshire'

After Dr Richard Richardson the elder died in 1741 Knowlton wrote occasionally to Richard, the younger, who as his father's heir inherited the estates at Bierley. These letters he directed to 'Mr. or Richard Richardson Esq.' Was it perhaps to this member of the family that Knowlton addressed the letter below, of 18 November 1739, when he directed it to 'Dr Richardson', thinking that the addressee had received the doctor's degree?

TO 'DR. RICHARDSON JUNIOR'

9ber [November] ye 18, 1739[(3)]
'worthy S^{r}.

I am very much obligd to you for y^{r} so ready answer — I wrot m^{r} Catesby word of ye same I am very sorrey to heare m^{r} Brewer is so very Il [ill] I have not heared from him since mickellmas but I hope hel recover... & beg youl be so good to give my most Humble Sarvice to him — I am very Glad to heare you have so grat a satisfaction in y^{r} stoves the Pines [*Ananas comosus*] are delicate frut & worthy all care &c as to y^{r} Jamaco Lillys I take it

(1) *Extracts from the literary and scientific correspondence of Richard Richardson*, etc. [Edited by Dawson Turner], xxxiv (1835).

(2) Smith (R. W. J.), *English-speaking students of medicine at the University of Leyden*, 195 (1932); Richardson is only mentioned in this work as a student at Leyden.

(3) Richardson corres. MS. RAD. TR. C. 10.

to be ye Lilly brought over by m^{r} Thourogood Super-cargo of ye Royale george from ye bay of Mixico in ye yeare 1722 or 3 & m^{r} Fairchild gott y^{m} who keep y^{m} till they flowerd in his garden & caused a print to be Ingraived of y^{m} & collourd of wch I had one or 2 of y^{m} —'

The grammar of the last few lines is confused. The present writer understands the meaning to be that the Hoxton nurseryman Thomas Fairchild obtained the bulbs from Mr Thourogood after the *Royal George* returned from the Gulf of Mexico. Fairchild planted some of these bulbs in his garden and when they flowered he had an engraving made of them. Subsequently, he gave 'one or 2 of y^{m}' to Knowlton, but was the 'them' referring to prints or bulbs? There appears to be no published record of such an engraving ever having been made.

A supercargo was a person in a merchant-ship who managed sales, etc. of cargo, and Mr Thourogood was probably travelling in a merchant-ship named *Royal George*. It is improbable that he was on the Ship of the Line of that name which was a very big naval vessel with one hundred guns, and most unlikely to be in service anywhere but in home waters.[1]

Later in this letter Knowlton returned again to the plant Mr Thourogood brought back from the Gulf of Mexico, but meanwhile he continued as follows:

'... y^{r} china rose or ketmia [probably *Hibiscus mutabilis* L.][2] I have not seen flower & is a rarity in its kind... ye above Lilly I had flower in ye spring plentyfuly & Likewise this autumn — Dr Chambers tels me ye Hortus Clifordiana [i.e. Linnaeus's *Hortus Cliffortianus,* published in Amsterdam in 1738] is not to be Bought he makeng onely preseants to his frainds'

It was presumably Knowlton's friend Dr Chambers of Hull who gave him the information about the *Hortus Cliffortianus.* This folio volume with the place of imprint and date Amsterdam 1737, was in reality not published until 1738. It consists mainly of a detailed catalogue by Carl Linnaeus of the plants in the gardens and herbarium of the rich Dutch banker George Clifford at his estate at Hartecamp between Haarlem and Leyden. Many of the engravings are made from drawings by the botanical artist George Dionysius Ehret (1708–70). Clifford, who

(1) Information supplied by Mr Pearsall of the National Maritime Museum.
(2) *See* above, p. 134.

George Dionysius Ehret (1708–1770), a German botanical artist of great distinction who settled in England in 1736. He illustrated Linnaeus's Hortus Cliffortianus *among other works.*

published the book at his own expense, did not wish to sell it, only to give copies away as presents.(1)

Knowlton closed his letter as follows:

'... my complements wait one y^{r} father & Mad^{m} Richardson & Dr Stanope & belive me S^{r} y^{r} most obligd & Humble sarvant
Tho Knowlton'

A postscript reads:

'I am Like to have a fine parsell of pine aples this next yeare have severall fine plants tho not so Large as yrs, hears very sharp weather for ye Green house plants & thos abroad & I am afraid m^{r} Brewer plants will sufer this hard weather if he is not well enought to Look after y^{m} pray tell him I sent a Lett^{r} by Thos foge who went from me into Lancashire & that I have had no answer to; & should be glad to heare how his garden furniture dos — the before mention'd Lilly was Brought from thence under ye name of flos Tigeritis'

The plant brought over in the ship returning from the Gulf of Mexico was possibly *Tigridia pavonia* (L.f.) DC., Tiger Flower, a bulbous perennial native of Mexico. This species early attracted the attention of the Spanish conquerors of Mexico, writes Molseed,(2) and became known by reputation under the name Tigridis flos (Tiger Flower) long before it was seen by any botanist. Several pre-Linnaean botanists described and figured the bulb under the name Tigridis flos, always with a note saying that it might be a species of *Dracunculus,* but none of these authors had seen the living plant. The first naturalist to see *Tigridia* was the physician Francisco Hernandez (1514–78) who had been sent by Philip II, King of Spain, to observe and describe the plants, animals, and minerals of Spanish America. He described and illustrated the plant under the heading 'De ocoloxochitl [the Aztec name meaning Tiger flower] seu flore tigris'.(3) The first true binomial name was given to it by Carl Linnaeus the younger, who called it *Ferraria pavonia.*

(1) *See* Linnaeus (C.), *Species plantarum, a facsimile of the first edition 1753,* printed for the Ray Society, **1**, 49 (1957), Introduction by W. T. Stearn.

(2) For a history of this plant *See* Molseed (E.), The genus *Tigridia* (Iridaceae) of Mexico and Central America, *University of California publications in botany,* **54**, 37–43 (1970).

(3) Hernandez (F.), *Rerum medicarum Novae Hispaniae,* 276 (1651) Rome.

In 1802 the plant was described by A. P. de Candolle, making the new combination *Tigridia pavonia* (L.f.) DC.,(1) the name by which it is known today. According to the *Botanical Magazine,* **177**, 544 (1969) '*Tigridia pavonia* seems to have been introduced into this country about 1795, for it is recorded as having flowered and set seeds with Mr. Ellis Hodgson of Everton [near Liverpool] in 1796.'(2) If the plant Knowlton had flowering in his garden in 1739 was *Tigridia pavonia* it is by far the earliest record of the plant being cultivated in this country. Unfortunately, we cannot be certain of this as he left no description. However, if the *Royal George* visited Mexico and called at the seaport of Vera Cruz the supercargo Mr Thourogood may well have seen some of the large and striking flowers of *Tigridia pavonia* and as a result decided to obtain some of the bulbs to take back to England. This would perhaps have not been very difficult as *Tigridia pavonia* occurs in pine-oak forests throughout Mexico and in the pine-fir zone of southern Mexico and Guatemala.(3) It is also possible that the nurseryman, Thomas Fairchild, noted as a grower of exotics, asked Mr Thourogood to get him some of the bulbs.

Before finally ending his letter Knowlton returned once again to the subject of his Pineapple plants:

'I have no courious plants in flower now & ye reason is I keep y^{m} in so small potts pervents y^{m} but am oblidge for to give rome [room] to my pine aples &c I bege youl give ye Inclosed to m^{r} Brewer & youl Infinatly oblige . . . y^{r} most obydant Humb^{le} sarvt
Tho Knowlton.'

Knowlton's rare plants were noted as a subject of interest by the Reverend William Stukeley in a letter, dated 1 July 1740, when, after a visit made to Londesborough with two friends, one of whom was the antiquary Francis Drake, he recounted:(4)

'. . . to Londsburrow, a seat of Lord Burlington's; the house much improved by my lord, the park elegant, the greenhouse very curious in innumerable, rare, and exotic plants.'

(1) In Redouté (P. J.), *Les liliacées,* **1**, plate 5 with descriptive text (1802).
(2) *See* also *Bot. Mag.,* **15**, 532 (1801).
(3) Molseed (E.), *op. cit.,* **54**, 37–43 (1970).
(4) Stukeley (Revd. William), *The family memoirs of the Rev. W. Stukeley*, etc. [Edited by W. C. Lukis], **3**, 381 (1887) Surtees Society **87**.

TO PETER COLLINSON

Londesborough, Yorkshire, Jan:20:1740/41[1]
'Sir,

Last Summer I made an Excursion into several of our neighbouring Counties. In this journey I observed the Misleto [*Viscum album* L.] growing plentifully on the common Maple [*Acer campestre* L.] at Walling Wells [about three miles north east of Worksop], and on the Oaks not far from thence likewise on the Abele's [i.e. *Populus alba* L., White Poplar, Abele] in an Avenue belonging to the Earl of Holderness's Park near Rotherham Yorkshire.

I have seen it on the Lime [possibly *Tilia* x *europaea* L., Common Lime] at Connons [i.e. Cannons, the Duke of Chandos's estate near Edgeware] and at Bongate [Bone Gate in East Barnet] in the same County [of Middlesex].[2] It grows with me on the Quickbeam on Sorbus sylvestris [*Sorbus aucuparia* L., commonly called Rowan or Mountain Ash, and, in Evelyn's *Sylva,* Quick Beam],[3] and on the Aria Theophrasti or whitebeam [*Sorbus aria* (L.) Crantz, White Beam] and on the Acacia Robeni [*Robinia pseudoacacia* L., Acacia]. I once found it on the Hazel [*Corylus avellana* L.] near Market Street in Hertfordshire.

As it is rare to find the Misleto on these Trees, but especially on the Oak, I hope it will not be unacceptable to acquaint you with it.

I am, Sir, Your humble Sarvant,
Tho: Knowlton.'

Peter Collinson was very interested in the subject of Knowlton's letter. He himself left a record of some observations he made in 1739 on the growth of mistletoe as the result of droppings from thrushes on the boughs of trees in Lord Petre's garden at Thorndon, in Essex. A list composed from notes received by Collinson recording the names of trees on which Knowlton had found mistletoe growing was published in *Hortus collinsonianus,* 57 (1843), edited by L. W. Dillwyn. Some of the entries differ from those that appear in Knowlton's letter of 20 January 1741. They are given below:

'...On Apple trees...
On the Pear tree
On the Maple in Yorkshire and Huntingdonshire...

(1) Letter preserved in the Library of the Royal Society (Letter Book LBC. 26, 220). It was read at a meeting of the Society on 5 March 1740/41.
(2) *See* John Rocque's map of 1754.
(3) *See* Hunter's edition (1776) of Evelyn's *Sylva,* 228.

On the Elm,
On the Willow,
On the Buckthorn,
On the Sallow,
On the Service,
} These all were observed at Esquire Blackburn's, and in Lancashire and in Westmorland, in 1764.
On a Holly Branch, which was at a druggist's in Bow-lane, Cheapside.
On a Virginia Walnut tree, growing in our fields at Mill Hill.[1]
On the Oak (which is very rare), Mr Knowlton has twice seen it. In August, 1765, three plants were found growing on the Oak in the estate of – White, Esq., at Watling Wells, [Walling Wells, near Worksop, Nottinghamshire]'

When the Earl of Burlington was not in residence at Londesborough, Henry Simpson, who appears to have been the Earl's personal secretary *cum factotum*, based mainly in London, corresponded with Knowlton regarding various matters, including those connected with the garden at Londesborough.[2] Below are given extracts from some of these letters, and it is interesting to note the high esteem with which Burlington held the Pineapples:

London July ye 11th 1741
'Sr Mr Knowlton

My Lord... Order'd me to tell you, to send some of the Pine Apples to Mr Gee, as they become ripe to send some often up to Town directed to his Lord^{p} and ye rest you may dispose on, to such of his Lord^{ps} friends as you shall see proper

I am
S^{r}
Yo^{rs} &c H.S.'

'Mr Gee' identified with James Gee (1686–1751) and generally known as Colonel Gee,[3] was educated at Queen's College, Cambridge, and the Inner Temple. In 1715, at the time of the Jacobite Rebellion, he was appointed Captain and Lieutenant-Colonel in the Third Foot Guards.

(1) Collinson lived at Ridgway House, Mill Hill.

(2) Corres. in the Devonshire collections at Chatsworth, Letter Book 42. L.i.

(3) For details of James Gee *See* Neave (D.), High and Low Halls, Bishop Burton, *The Georgian Society for East Yorkshire,* Issue No. 6 (1979). Previous to this publication the Colonel Gee who had a design made for a house at Bishop Burton had usually been identified with Colonel William Gee (1699–1745).

Colonel Gee lived at Low Hall, situated to the south of the church in Bishop Burton, near Beverely, in a house possibly designed by the Earl of Burlington. He was uncle of Sir Charles Hotham of South Dalton, and a close friend and a relation through marriage of the Earl. From correspondence preserved at Chatsworth it is learnt that in October and November 1739 Gee was a frequent guest at Londesborough, and during the same period Burlington dined once at Bishop Burton.

London July ye 25th 1741
'S^{r}: Mr Knowlton
I have My Lords orders to acquaint you that this day Mr Arundell & his Lady sets forward for Yorkshire & go to his house where they stay but one night & from thence proceed to Scarbrough & intend to take Londesburgh in the way thither wch if they shou'd his Lord^{p} desires they may have fresh fish out of ye ponds in the garden Pine Apples or any sort of fruit or anything produced in the garden or under your care or Managem.^{t} and at same time you are to ask him what Fish, Pine Apples, Fruit, or what else he wou'd be pleas'd to have sent to Scarbrough... if he shou'd not call there you must write to him at Scarbrough and desire to know what... he wou'd be pleas'd to have sent to him there... I am

S.^{r}

Yo.^{rs} &c HS.'

This was the Hon. Richard Arundell (*c.* 1696–1758) of Allerton Mauleverer, north west of York, in the West Riding. He was Member of Parliament for Knaresborough 1720–58, and in 1732 he married Lady Frances Manners. He was an intimate friend of the Cavendish family.

London Augt ye 8:1741
'S^{r}:Mr Knowlton
I reced y.^{r} Łre giving Advice of y.^{r} having sent 8 more Pine Apples to My Lord & I sent y^{r} Łre to Chiswick to My Lord since wch ye Box w.th ye Pine Apples came safe and I sent 'em therein to Chiswick also but have since heard nothing of 'em.

I am

S.^{r}

Yo^{rs} &c HS'

London Augt ye 8:1741
'Sr Mr Knowlton
I have writ before to you by this post and have sent it away since wch I have receiv'd a Łre from My Lord who is at Chiswick commanding me to write you by this post that he wou'd have you come up to Town for about a week wch I desire you will do imediately on Receipt of this.

I am
S^{r}
Yo^{re} &c HS.

PS
I desire youl give me Notice of your comeing'

London Jan:the 30:1741/42
'S^{r} M^{r}:Knowlton
I have reced your Łre of the 27. Inst. & this day shew'd ye same to my Lord, who Order'd me to write you to plant again that part of the Avenue you mention in the Crossthorns, and... other places where you find necessary...

I am
S^{r}
Yo^{rs} &c HS.'

London July ye 3^{d} 1742
'Sr Mr Knowlton
reced y^{r} Łre of the 29th past & Yesterday lay ye same before My Lord who order'd me to write you to send Pine Apples hither as they shall become ripe and if there happens to be more than you can send hither his Lord^{p} Orders that you give 'em to such Gent in the Countrey as you know to be his friends I am

S^{r} Yo^{rs} &c HS'

London July ye 17:1742
'S^{r} M^{r}: Knowlton
Yo^{rs} of the 11:Inst came safe to hand... & Thursday Evening the Pine Apples arriv'd 4 of wch I took out by My Lords Order and sent to my Lady Dow^{r} in Pall Mall(1) wth whom his Lord^{p} My Lady:Lady Charlotte(2) &c Din'd Yesterday who eat of the Pine Apples & comended em very much...

I am
S^{r}
Yo^{rs} &c HS.

PS:
I shall take care of ye Crowns of the Pine Apples as much as I can & desire youl let me know whether or no you wou'd have 'em sent down'

(1) Juliana, Countess of Burlington.
(2) The Earl of Burlington's youngest daughter, Charlotte.

London Augt the 26:1742
'Sr Mr Knowlton
I have read y^{r} two Łres one bringing the Monthly Acco^{t} of labourers to 7th: Inst the other advising of a box of Pine Apples sent by ye York Coach wth a small parcel of Coffee therein The Box came safe & I sent ye same to Chiswick

I am
S^{r}
Yo^{rs} &c HS.'

Some years earlier Knowlton had addressed a letter, dated 27 August 1736, to Brewer in which he told his friend that he had 'grat plenty of young coffe trees' (*Coffea arabica* L.)[1] By 1742, these trees would be bearing plenty of deep red fleshy berries, each containing two seeds – the coffee beans of commerce. It may be assumed, therefore, that the small parcel of coffee received by Burlington consisted of coffee beans from the trees grown on his own estate.

London Octr the 13:1742
'S^{r}: Mr Knowlton
I reced y^{r} Łre . . . My Lord . . . Order'd me to write you by this post to send by the York Coach what Pine Apples you have ripe for this time only; and no more till further Orders from his Lord^{p}. . .

I am
S^{r}
Yo^{rs} &c HS'

London Octr the 23^{d} 1742
'S^{r} Mr Knowlton
I have My Lords Orders to write you by this post that as he will shortly be down in the North he would have you imediately provide yourself wth mourning vizt a dark grey Frock & a black Wastcoate & Breeches & his Lordp gave no further Orders

I am
S^{r}
Yo^{rs} &c H.S.'

For whom mourning dress was required does not seem to be known. Ralph Wise, Burlington's agent at Londesborough, was sent a similar

(1) *See* above, p. 47.

letter requiring him to provide himself 'imediately wth Mourning'. Earlier in the year, on 2 May, there occurred the death of Burlington's daughter, Dorothy, Countess of Euston, and if this was his first visit north since then it is possible mourning was required for her.

London Octr the 28: 1742
'S^{r}: Mr Knowlton,
I have My Lords Orders to write you that he intends to set out for Londesburgh next Tuesday wth the Lady Isabelle Finch & goes to Lord Maltons where he intends to be on Saturday sennight & will stay there all Sunday & on Monday Morning set forward for Doncaster & from thence proceeds to Londesburgh to Dinner where he intends to be on Monday sennight being y^{e}: 8th of November next... his Lordp. will stay about a fortnight I am
S^{r} Yo^{rs} &c HS.'

Thomas Watson-Wentworth, Earl of Malton (1693–1750) married in 1716 Mary, daughter of Daniel Finch, seventh Earl of Rockingham. He had vast estates in Yorkshire, and Burlington would have stayed with him at Wentworth-Woodhouse, in the West Riding, before continuing north to Doncaster.

London July the 9th 1743
'S^{r}: Mr Knowlton
I reced y^{r} Ł re of the 1: Inst & shewd y^{e}: same to My Lord who order'd me to write to you, to send up hither six Pine Apples to his Lord[ship] and if you have more... to send to the Archbp of York... & if... you have any spare he woud have 'em given to... his friends, as you have usually given to
I am
S^{r}
Yo^{rs} &c HS.'

TO SAMUEL BREWER

Londesburght Xber [December] ye 15 [1741][(1)]
'D^{r} S^{r}
I am but just come home[(2)] & gott a Little rest after my fatague... just before I Left yorkshire I sent you a Box of plants & within ye same a Little

(1) Knowlton corres. British Museum (Natural History). Xber, i.e. the tenth month (December) of the Roman, or Julian Calendar. *See* p. 151.
(2) Knowlton had apparently just returned from one of his occasional visits to London. For example, in August 1741, at Burlington's request, he was asked 'to come up to town for about a week.'.

parsell of seeds in paper & ... a present of a geney [guinea] wch I bege y^{r} aceptance wch I hope come safe to y^{r} hands ... m^{r} Collison Likewise sent you another Inclosed in a Letter freed by ye Duk of Richmond[1] ... he wonderd much it shoald not be taken notice of as realy I did ye same neither of us wanted ... Thanks for so small a Tryfel but ye frank accknowlegement of em

in my way down was a whitthorn tree in as full Blow as if in ye midle of may it was growing by ye roadside neare ye Turn Pike 2 or 3 milles beyong Stillton [i.e. Stilton, a few miles south of Huntingdon] towards London so that all Travellirs pulld & fild thire hatts & Brest with ye same ...'

The tree noted by Knowlton, was probably *Crataegus monogyna* Jacq., popularly known as Hawthorn, Whitehorn, or May.[2] The road by the side of which the tree was growing must have been the Great North Road, on part of which Knowlton would have travelled from Londesborough to London. Stilton was one of the stages on this road. Professor H. H. Lamb of the Climatic Research Unit of the University of East Anglia has told the present writer:[3]

'... that the explanation of the abnormal flowering of the Whitethorn seen by Knowlton seems to be that it was a very warm Autumn, possibly right up to and including the first week of December, after a warmish Summer; but the winter that followed soon turned cold.'

Knowlton recounted to Brewer:

'Ld Wilminton & Pittrea [Lord Petre] have fine collections of plants I have Brought some few along with me so hope shall perserve y^{m} some being curious old ones but I found many good ones Lost ...'

The first of these peers has not previously been mentioned by Knowlton. He was Spencer Compton, Earl of Wilmington (*c.* 1674–1743), the owner of estates in Sussex, including Compton Place at Eastbourne, where there was a garden designed by the King's gardener, Charles Bridgman (d. 1738). In the preface to the *Catalogus plantarum* of 1730,

(1) At this time letters from members of both Houses of Parliament were carried free of charge in the post.

(2) *See* Grigson (G.), *The Englishman's flora,* 166 (1955).

(3) In *epist.* 11.iv.78.

published by a Society of Gardeners, the work done towards the advancement of horticulture by Lord Wilmington and others is acknowledged as follows:

'Nor should we in mentioning particular persons forget the many curious gentlemen which at present are carrying this spirit of gardening to a considerable height, by introducing many new kinds of plants, flowers, trees and fruits, and in making many curious experiments concerning their culture and uses...'

At the time Philip Miller published his *Gardeners dictionary* in 1731, Lord Wilmington's gardener was Lewis Kennedy.(1) This may have been the person of that name who was gardener at Woburn to the Duke of Bedford, and in 1743 committed suicide.(2)

Knowlton continued his letter by giving his friend further news of his visit south:

'Sr Hans Slone [Sloane] Mr Collison [Collinson] Ld Pittrea [Petre] Mr Rand [the 'Praefectus horti and Demonstrator of plants' at the Chelsea physic garden] & all ye other frainds are well & as to Miller [Philip Miller, gardener at the Chelsea physic garden] I deliverd y^r^ compliments to him but I do asure you hel not Look one [on] a poore man; nothing Lesse then a L^d^ or a duke is Company for him so if you take my advice never trouble ye Coxcombe with any thing more; I have said all I could in y^r^ faver...

All ye Rest of y^r^ acquanteence are good in asking after your health in a most fraindly way... & are willing & ready to sarve you – I dined one day with Mrs Green & Mad^m^ Sherrards [the widow of James Sherard of Eltham in Kent, where Knowlton was a gardener as a young man]... ye Latter so much a child now it gave me a Little uneasynesse to hear of it... Mrs Green was saying she was sorrey y^t^ you had not some of ye plants who was so great a Lover Indeed they was goven to any as would have y^m^ as I have heard(3) – James Gordon who was his [i.e. James Sherard's] gard^r^ after Sampson & since with L^d^ pettrea is now setled att Mile End has Bult a stove with his own hands & has pickt up a good many od things perhaps not easely to be mett with & sells seeds &... abounding with more Knowlege y^n^ 500 Millers of Chelsey putt togather &c

Mr Christopher Grey of fullham has Published his Cat[alogue] of his

(1) *See* List of subscribers.

(2) *See Gent. Mag.*, **13**, 543–44 (1743).

(3) *See* also above, p. 175.

garden wch chefely consists of great verrity of shrubs many of wch are new and as hardy as any of ye old & stand well in ye open are as oaks, as do ye Tackamahack in many plases [i.e. the tree named *Populus tacamahaca* by Miller and *Populus balsamifera* by Linnaeus, commonly called Balsam Poplar, Taccamahac, native of North America] . . .[1]'

In the above passage Knowlton may have been referring to *A catalogue of American trees and shrubs that will endure the climate of England.* Although undated, this rare single sheet was probably issued about 1737–40. It lists the trees and other plants 'to be had of Christopher Gray Nursery Man at Fulham near London', and contains a fine engraving of 'Magnolia altissima' (*Magnolia grandiflora* L.), signed with Mark Catesby's monogram.[2] Knowlton continued:

'. . . ye Spining Youcas [probably *Yucca filamentosa* L, Silk Grass, native of America] are att my Ld Ileas growing in ye open grownd in ye finest health I ever see plants as do many more things of y^{t} kind . . .'

It appears that during his visit south Knowlton had been to Lord Ilay's property. This was Arthur Campbell, Earl of Ilay (1682–1761), who in October 1743 succeeded as Duke of Argyll. He made his gardens at Whitton Place, near Twickenham, Middlesex, famous for their fine collection of exotic trees and shrubs.

Knowlton also told Brewer:

'. . . ye Hyacinthus afr[icanus] umbelifera flore ceralus Tuberosus radice [*Agapanthus africanus* (L.) Hoffsgg., African Blue Lily, native of South Africa] . . . is a plant almost Lost about town & stramonium flore duplic & triplic albo a curiosity I had never seen before this jorney [possibly double and triple forms of *Datura metel* L., Horn of Plenty, native of India] – I hope . . . you & ye worthy Drs Richardson & Stanope [are] all very well & bege youl be so good to present my most Humble sarvce unto y^{m} & . . . I wish y^{m} & you a merry Xmis & many Hapy new years . . . hope to see y^{m} att ye Election to vote for a fox wch I shall do & hope many of our nabiors for it would be ye most hat[e]full thing in ye world for me to joyn with pensioners & Place men &c . . .'

(1) Knowlton had seen this species growing in Guernsey, *See* above, p. 57.

(2) *See* plate facing p. 348 in Henrey (B.), *British botanical and horticultural literature before 1800*, **2**, (1975). Mark Catesby adapted this engraving from plate 61 in vol. **2** of his *Natural history of Carolina,* etc.

On 8 May 1734 Dr Richardson's son-in-law Sir John Lister Kaye, Bart. had been returned as a Tory for the constituency of York, but he did not retain the seat in the election of 13 May 1741. In 1742 he was invited to stand again, but he refused on grounds of health, recommending another wealthy Tory, George Fox of Bramham Park, Yorkshire, who was returned unopposed. The election took place on 21 July 1742.[(1)] Knowlton appears at that time to have been very politically minded and anti-government. He therefore referred contemptuously to pensioners and placemen who were supporters of the Whig government.

Knowlton ends his letter, thus:

'Mr Catesby is very well & sends his sarvce to all his fraindes & will very soon publish his Last part to compleat ye 2 voll.^{m}[(2)] wharein youl have a map of Caralina with a Long Disertataion one [on] Birds of ye East & West Indies &c with severall of ye most curiosus plants in America so bege youl recom̄end it to y^r fraindes & belive me S^r y^r most obydient Humb^{le} sarvant...

Tho Knowlton

PS you shall be welcome in spring to any plants I have or can spare so I bid you welcome &c pray Lett me Know when Dr Delenus [i.e. Dillenius] History of moses [mosses] comes out [published in 1742][(3)] & what is ye Best news from Oxford I shall be glad to heare how you do & what new plants you have and when you heard Last from Mr Blackburn he corespondes with Mr Gorden [i.e. James Gordon, the nurseryman] & had ye Digitalis accanthoides [*Isoplexis canariensis* (L.) Loudon, native of Canary Islands] ... since I went up to London ... pray Lett me heare how Dr Richardson stoves & gardens gos one [on] I long to see you & y^m &c'

When he wrote this letter it appears that Knowlton did not know that Dr Richard Richardson had died the previous 21 April.

Knowlton's invitation to his friend to visit him at Londesborough was accepted, and in a letter of 2 June 1742 written by Peter Collinson to Samuel Brewer, we read:[(4)]

'I am glad you have been able to take a Turn to see Mr Knowlton I

(1) Sedgwick (R.), *History of Parliament, The House of Commons 1715–1754*, **1**, Introductory survey, appendices, constituencies, members, A–D, 364 (1970).

(2) i.e. the tenth, and last part, of Catesby's *Natural history of Carolina, Florida,* etc., which was published in 1743. The *Appendix* appeared in 1747.

(3) Title page dated 1741.

(4) Collinson MS. at end of Knowlton's corres. in British Museum (Natural History).

Brass flagon, with detail of cover, drawn by Knowlton in a letter to Sir Hans Sloane in 1741. It was found in ploughing near the city of York, and was later illustrated in Camden's Britannia.

doubt not but that you was well entertain'd – I know He has a great Respect for you.'

Not many days after he wrote to Brewer, Knowlton sent a letter, dated 27 December 1741, to Sir Hans Sloane, in which he gave details concerning certain occurrences he considered unusual and interesting.

In this letter, beneath a drawing of a medieval tripodial, globular jug, with a cover in the form of a male bearded mask, on its forehead a figure, part bird, part animal, sitting, intended as a handle, Knowlton wrote:

TO SIR HANS SLOANE

27 Xber [December] 1741[1]

'...the form of an old Brass flagion found neare york in plowing up som grownd very Lately 1741 the waight of ye whole is 17 pds[2] 4 ounces & half the Lid alone – 3 pds 6 [oz] & holds 5 pinnts one gill & a half
Most worthy & Hon^{rd} S^r
the above sketch is ye flaging I send you as a curiosity of great antiquity wch is thought so heare & hope will not be unexceptable to you...'

In Camden's *Britannia,* **3**, 304, pl. 16, fig. 13 (1806) is the figure of a flagon found in 'ploughing near the city' of York identical with Knowlton's drawing. The weight of this flagon, the lid alone, and the

(1) Sloane corres. 4057 fol. 100^a–101^a in British Library.
(2) The weight, although difficult to read, appears to be 17 lbs.

measure of liquid capacity, is also clearly the same as that mentioned in the gardener's letter. It seems reasonable to assume, therefore, that the flagon found in Yorkshire in 1741 is that mentioned in Camden's *Britannia,* although no date is given there for the finding.

In this letter the gardener repeated to Sloane the story he had told Brewer about the Whitethorn Tree he had seen in bloom on the road to London:[1]

> 'Xber [December] ye 4.th 3 milles south of Stillton... I see a white thorn tree in... full flower as ever I see one in ye midle of may... many Hunderid of people or most who passed by one [on] ye Road gatherd & pluck[ed] such quantitys to adorn there hatts & Bussomes as I myself did to my very great Surprise & about a week before was in M^{r} [Christopher] grays nursery could not see one of his Glasenburry thorns so much as offering to bud.'

In the south and west of England if the weather is mild the Glastonbury Thorn, *Crataegus monogyna* Jacq. cv. Biflora, sometimes flowers and produces young foliage in winter.[2] It may be noted that Christopher Gray's nursery was situated at Fulham, in the Thames Valley, where plants usually come into flower earlier than in many other parts of the country.

Knowlton ended his letter with a passage in which he described another spectacle that had interested him. He wrote:

> 'Neare Lincolon [Lincoln] I see severall of ye Largest of sort of sea gulls haveing Black wings onely; ye rest of ye Body all white a bird very raire in our parts... from y^{r} Hon^{rs} most obydant Humble sarv^{t}
> *Tho Knowlton*'

When he travelled from Londesborough to London Knowlton probably went through Market Weighton, Beverley, across the Humber to Hull, and then to Lincoln before continuing his journey from Newark on the Great North Road. The birds seen were probably the Great black-backed Gull, *Larus marinus.* Thomas Pennant in his *British Zoology* of 1766 (vol. **2**,

(1) *See* above, p. 196.

(2) Bean (W. J.), *Trees and shrubs hardy in the British Isles,* **1**, 777 (1970). *See* also Vickery (A. R.), *Holy thorn of Glastonbury,* (1979), West Country Folklore No. 12.

on Birds), in writing of this species stated 'inhabits our coasts, in small numbers and breeds in the higher cliffs.'

In a postscript the gardener added:

'...pray S^{r} when you see Mr Catesby & Mr Edwards present my Humble Sarvice unto y^{m} with a merrey Xmis & many hapy years be thers & y^{r} attendants'

'Mr Edwards' was the artist and naturalist George Edwards (1694–1773), author of several books on birds and other subjects.

The above letter to Sloane was read at a meeting of the Royal Society on 7 January 1741/42. Among other letters Knowlton addressed to the President was one dated 5 February 1739/40. This was read at a meeting held on 14 February and it:

'...served to communicate several draughts, which he [Knowlton] caused to be taken, of the appearance of a frost upon glass windows, and which resemble the forms of various flowers, herbs, trees &c.'

After Dr Richard Richardson's death in 1741, Knowlton corresponded with his son Richard Richardson the younger, who succeeded to the family estates and lived at Bierley. Knowlton sent the new owner of Bierley plants, and wrote to him on various matters concerning gardening and natural history. He also corresponded with his gardener William Stonehouse. Richardson was a Justice of the Peace and Deputy-Lieutenant for the West Riding. He was a collector of pictures and also of coins and made large additions to his father's library.[1]

TO RICHARD RICHARDSON, ESQ.

Tusday morn June 8 1743[2]
'D^{r} S^{r}

one [on] ye other side is a Cat: of 10 sorts of Plants in a Box Directed for you & in ye same...a Parsell of Plants in a Bundall for Mr Brewer with a Lett^{r} & one more Large Plant att ye Bottom [of] ye box he wrot for Last

(1) *Extracts from the literary and scientific correspondence of Richard Richardson, M.D., F.R.S.,* etc. [Edited by Dawson Turner], xxxiii (1935).
(2) Richardson corres. MS. RAD. TR.C.11.

month ... y[e] Treachellium curuelium umbeliferum [*Trachelium caeruleum* L., Blue Throatwort,(1) native of the western Mediterranean and Portugal] ... w[ch] I bege youl be so good as give unto him They was sent yesterday by y[e] Hull coatch to York & I hope youl Recd y[m] safe and in good order my complements wait on y[r] good Ladye with greatfull accknowlegement for favers &c & am S[r] y[r] most obedant Humble Sarv[t]
Tho Knowlton'

The various species listed in the Catalogue include the following:

'Euphorbium verum antiquorum [*Euphorbia antiquorum* L., native of India and Ceylon]
Double nasturtium [*Tropaeolum majus*]
Mimosa in M[r] Brewes Parsell...
Sawruras 2[d] in miller [i.e. in Miller's *Gardeners dictionary,* Ed. 6 (1752), Entry 2, Saururus racemosus, seu botryites minor, Lesser branching Lizard's-tail, now included in *Piper amalago* L.]
ficcoides [Ficoides, or Fig Marigold, *Mesembryanthemum*]
Annagallis monella [*Anagallis monelli* L., native of south-west Europe]...
Hernandiae [*Hernandia sonora* L., Jack-in-a-Box, native of West Indies and tropical America]
Stramonium Duplic flore &c [a double form of *Datura metel* L., native of India and Ceylon]...
Treachellium [*Trachelium caeruleum* L., Blue Throatwort] att y[e] bottom of y[e] Box with all y[e] Earth to it for M[r] Brewer...'

After Knowlton sent Richard Richardson the above mentioned letter and bundle of plants for Samuel Brewer, correspondence between Knowlton and Brewer ceased.

It will be remembered that Knowlton wrote to Brewer in February 1729/30 informing him that he had just made as many canals in Londesborough Park as Richard Richardson had at Bierley.(2) Moreover, that he was then in the process of constructing two more and when completed there would be a fine chain of them. When the ground for the largest of these canals, also known as the Great Pond, was being excavated the foundations of a Roman road were found. Knowlton informed the antiquarian Francis Drake (1696–1771) of the discovery and in his

(1) Sometimes commonly called 'Umbelliferous Throatwort'. *See A catalogue of plants in the garden of John Blackburne*, etc., 65 (1779).
(2) *See* above, p. 121.

Roman Road to York, 17 Miles from Garrowby Hill.
Roman Road to Malton, the CAMVLODVNVM of Ptolemy.
These Works continued and turn towards Burlington.
Garrowby Hill
The great Vale of York
Road to Pocklington
Roman Road to Surby and Burlington Bay, the SINVS SALVTARIS vel PORTVOSVS of Ptolemy
These Works continued towards Frydaythorp.
Road to Pocklington
These Works continued towards Whetwang.
DELGOVITIA
Millington
Huggit
Towards Hawood.
Kilwick
Road to Pocklington
This Line points to York Minster.
Scale of one Mile.
Wartre Hall
Low Wartre
High Wartre
Nunburnham
Roman Road
Road to Burnby
Road to North Dalton
Londesburg
Some Remains of a decayed Village
This Vale opens towards Bainton.
Kipling-Cotes Horse Race
Shipton
Godmansham
Wighton
To Brough
Road to Beverly
Roman Road to Beverley, PETVARIA, And to PRAETORIVM, Patrington, or Spurnhead: the OCELLVM PROMONTORIVM of Ptolemy.
This Line points to Beverly Minster.
PRAETORIVM Usque.
DELGOVITIA, M.P. XIII.
PRAETORIVM, M.P. XXV. Patrington, or Spurnhead.
NORTH RIDING
EAST RIDING
THE OCEAN
DELGOVITIA
PART OF LINCOLN SHIRE
Impensis Nobiliss. Dni Dni RICHARDI Comitis BURLINGTON

Eboracum: or, the history and antiquities of the City of York, **32** (1736), this writer recorded the following details:

'I shall beg leave to give the substance of two letters, which I received in answer to some queries, from Mr. Knowlton the noble lord's chief gardiner at that place [Londesborough]; a sensible, intelligent and a most creditable person. It is remarked that the road from Brough to Londesburgh park pail, is in a continued straight line; that it was formerly, and is still by some elderly people, called Humber-street; that the *stratum* of the road may be traced, under hedges, &c. across one of the canals in the park, which being lately made, occasioned the accident of finding of it. It is composed of materials very scarce in that country, and lies buried under a fine soil about fifteen inches; and it was with great difficulty that the workmen could dig through the agger. The curiosity of finding such a road in such an uncommon place, led my correspondent to trace it on both sides of the canals up the hills; and he can now, he says, shew it at any time with spades, one way pointing directly to the aforesaid Humber-street; the other up the park again, through that part called the Lawn, abutting up against hedges, trees, &c. clear to the Wolds; where it pointed either to Wartyr [i.e. Warter] or Nunburham [i.e. Nunburnholme], but which he had not then leisure to trace. The Malton and York roads lying that way...'

This was the Roman road that went northwards from Brough to Malton, passing through Warter on the way.(1) Lord Burlington had a map made, showing its route through Londesborough Park and entitled 'An accurate survey of some stupendous remains of Roman antiquity on the Wolds in Yorkshire... Through which some grand military ways to several eminent stations are traced,' etc. It was 'Survey'd and drawn by John Haynes of York. 1744.' and the engraver of the map was George Vertue.(2)

Some years after the discovery of this road important Roman remains

(1) For details of the route of the Brough–Malton road *See* Clark (M. K.), *A gazetteer of Roman remains in East Yorkshire*, 37–8 (1935) Roman Antiquities Committee of the Yorkshire Archaeological Society. At the end of the work is a map of 'Roman Malton and district.'.

(2) Copy in the Royal Geographical Society, London.

Opposite:

'An accurate survey of some stupendous remains of Roman antiquity on the Wolds in Yorkshire', surveyed and drawn by John Haynes, 1744. The Roman road is seen passing through Londesborough near the bottom of the map.

were found a few miles to the north of Londesborough, on a site on the north bank of Millington Beck, just to the west of the line of the Roman road. In September 1745, Knowlton wrote about this archaeological discovery to Mark Catesby who was a Fellow of the Royal Society. Subsequently, an extract from this letter was read at a meeting of the Royal Society held on 6 March 1745/46 and published in the *Philosophical Transactions,* **44**, 100–101 (1748 for 1746). A quotation is given below:

'Sir

Lately we have discovered (within Four Miles of this Place) many Foundations in a ploughed Field... It was discovered by one Mr. *Hudson*, a Farmer at *Millington,* as he formerly tended his Sheep on one Side of the Hill, and on the opposite Side had perceived in the Corn a Difference in Colour for some Years before; which led him this Summer to dig; and, happening on the Foundations, it encouraged him, Dr. *Burton,* and myself, to dig likewise in several Places; and in one Part was discovered a circular Foundation five Feet wide, and the Plan within 45 Feet Diameter; by which I imagine it was a Temple dedicated to *Diana*... there were likewise many other Foundations which had... *Roman* Pavements within them; by which I imagine, after the Dissolution of the Temple it became a *Roman* Station, then called Delgovicia; which has been so un-certainly fixed at *Goodmanham, Londesbrough, Hayton,* &c... the Ruins and Foundations are a Demonstration of the once Grandeur of the Place; and no doubt it was the above-mentioned *Delgovicia.* The Foundations lay about 18 inches below the Surface, and to the Depth of 4 or 5 Feet within the Ground.
N.B. It is on the *Wolds* Hills, within two Miles of *Pocklington.*'[1]

At a meeting of the Royal Society, held on 28 May 1747, was read 'A dissertation on the situation of the ancient Roman Station of Delgovitia in Yorkshire; by John Burton of York, M.D.' This was published in the *Philosophical Transactions,* **44**, 541–552, 7 pls., Appendix 553–556 (1748 for 1747), with 'An Appendix... by Mr. Fr. Drake F.R.S.' These two writers, John Burton (1697–1771) and Francis Drake, both eminent Yorkshire antiquarians, argued for the acceptance of Millington as the site of Delgovicia. They pointed out that the several buildings, including a circular stone building resembling a temple, two oblong buildings, fragments of tessellated pavements, Roman pottery, tiles, and coins, together with the situation of the site (which was near important roads), the availability of

(1) Millington, Pocklington, Hayton, and Goodmanham, are all in East Riding, within a few miles of Londesborough.

water, and the nature of the surrounding countryside, were all reasons for accepting Millington as the Roman station of Delgovicia, so long sought after by Camden and other writers.(1) However, as is shown below this opinion is not upheld by modern authors.

The Millington site continued to be examined during the nineteenth century and in 1935 Mary Kitson Clark published a work giving an account of the excavations as well as a chronological list of records.(2) Subsequently, Neil Loughlin and Keith Miller, authors of *A survey of archaeological sites in Humberside,* 119 (1979) pointed out that although Millington was said to be the supposed site of a Roman temple, 'in view of its structures and location, and by analogy with other sites, it is more likely to be a villa (and not a major settlement, as it has often been considered)'. Earlier, the same theory had been put forward by Clark.(3) Furthermore, in their book entitled *The place-names of Roman Britain,* 155–157, 331–332 (1979), A. L. F. Rivet and Colin Smith published part of the journey of the Antonine crossing to Britain by the Gauls. The part of the route shown is from Bremenio in the north to Praetorio on the river Humber, Bremenio being identified as High Rochester. Rochester in Northumberland now occupies the site of the Roman station. The three stages after York namely, Derventione, Delgovicia (variant Delgovitia) and Praetorio are in the opinion of Rivet and Smith identified as Malton 20 miles north east of York, Wetwang 13 miles south east of Malton, and Brough 22 miles south of Wetwang. The fact that 'there are no significant Roman remains at Millington' is one of the reasons given by Rivet and Smith for the introduction of the possibility that Delgovicia is not Millington but Wetwang.

Another letter by Thomas Knowlton addressed to Mark Catesby was read at a meeting of the Fellows of the Royal Society on 13 March 1745/46. This was published in the *Philosophical Transactions,* **44**, 124–127 (1748 for 1746) under the heading: 'An account of two extraordinary deers horns, found underground in different parts of Yorkshire'. Accompanying this 'Account' is a plate containing figures of the skull and horns of two species of deer. The first figure shows the skull and horns of a Red Deer found in about 1742 at Ness, eight miles north west of Malton, in Yorkshire, 'in a

(1) *See* [*William*] *Camdens Britannia ... with ... additions ... by Edmund Gibson,* Col. 737 (1695), etc.

(2) *A gazeteer of Roman remains in East Yorkshire. Roman Malton and district Report No. 5,* 109–111 (1935). Published by the Roman Antiquities Committee of the Yorkshire Archaeological Society.

(3) In *A gazeteer,* etc., 58 (1935).

sand-bed, in the River Rye, which runs into the Derwent, in the East-Riding'. This skull may, or may not, have been fossil. It is not an uncommon find as this species of deer *Cervus elaphus* is found alive in the British Isles at the present day. The second figure is of considerable interest as it is an illustration of the skull and horns of the Giant Deer, or 'Irish Elk', *Megaloceros giganteus*, a species that became extinct about 10 000 years ago. The skull illustrated 'was found in a Peat-moss, at Cowthrop near North Dreighton in Yorkshire, in the year 1744.' This was an unusual find and is probably the first of this species to be found in England.[(1)] It was possibly from a late glacial deposit under peat, as in more recent finds in northern England and Ireland.[(2)]

At the end of Knowlton's letter of September 1745, mentioned earlier, addressed to Mark Catesby, and read at a meeting of the Royal Society 6 March 1745/46,[(3)] Knowlton gave details of the large monolith in the churchyard of All Saints, Rudston, about five miles west of Bridlington in the East Riding. Another of his letters read on the same occasion was 'of two men of an extraordinary bulk and weight' who lived in the neighbourhood of Halifax in West Riding. This letter is dated 28 February 1745/46.

On 11 May 1749, at a meeting of the Royal Society, Knowlton was introduced to the Fellows by the Secretary, Dr Cromwell Mortimer. He was evidently referring to some of the persons he met on that occasion in the following letter of 18 July 1749 addressed to Richard Richardson:

TO RICHARD RICHARDSON, ESQ.

July 18th 1749[(4)]
'Dear Sir,

When I was in London I saw Sir Hans Sloane, Drs Mead, Stack & Mortimer[(5)] & was very frequently with them; from whom I recd several

(1) The present writer is indebted for assistance in the identification, and details, regarding this species, to Dr Adrian Lister.

(2) *See*, for example, Mitchell (G. F.) and Parkes (H. M.), The Giant Deer in Ireland, *Proc. R. Ir. Acad.*, **52B**, 291–314 (1949).

(3) Published in *Phil. Trans.*, **44**, 101–102 (1748 for 1746).

(4) Richardson corres. MS. RAD. TR.C.11. This letter is not in Knowlton's hand, and as it is in more conventional English, it was presumably composed from Knowlton's dictation.

(5) Sir Hans Sloane, President of the Royal Society, and Richard Mead, Thomas Stack, and Cromwell Mortimer, Fellows of the Royal Society. All four men were eminent in the field of medicine.

seeds & other Favours; I saw likewise Messrs Catesby & [George] Edwards who has materials for a 3.^d Volume of Birds, Flies & Animals &c. but poor Mr Catesby's Legs swell & he looks badly, D.^rrs Mead & Stack said there were little hopes of him long on this side the Grave; he spoke to me of his Appendix [to *The natural history of Carolina,* etc. (1747)] which he said you had not got; therefore desired me to speak to you about it... there are... many Plates: you'll see the Proposals which I happen'd to have one of so send it in the Box of Plants... if you please to have it...

The infinity of new Trees, Shrubs &c. now of late introduced by him into the Gardens from North America fill me with the greatest Wonder & Astonishment imaginable. From Dr Meads favours I have rais'd the Azaderach [*Melia azedarach* L., Bead Tree, native probably of India but early introduced to China and possibly from thence to Europe] which is a rare & scarce Plant; as I have several others: I am much obliged to the D.^rs great good Nature to me in many Particulars. I send you the following List of plants which I hope will come safe... who am Sir Your obed.^t humble Servt *Tho Knowlton.*
P.S. The Box was this Day sent to the Carrier...'

Among the plants listed and sent to Brewer were the following:

'Arum Aegyptiacum [*Colocasia esculenta* (L.) Schott, native of India and Ceylon, commonly called Taro]...
Anagallis monella [*Anagallis monelli* L., native of south-west Europe]...
The new Strawberry [*Fragaria chiloensis* (L.) Duchesne. This strawberry was introduced into France from Chile in 1714 by a Frenchman, Amédée François Frézier, and in 1727 Philip Miller brought some of the plants back with him from Holland which had been given him by Gerard George Clifford who had large beds of them growing in his gardens at Hartecamp. J. J. Dillenius included a description and an illustration of a flowering specimen in his *Hortus elthamensis,* **1**, 145–146, pl. 120 (1732), made from a plant that bloomed in 1730 in the garden at Eltham. The introduction of this strawberry was most important historically as it became one of the parents of the modern large-flowered and large-fruited cultivated or garden strawberry][(1)]
Papay 2 Plants [*Carica papaya* L., Papaya, native of the American tropics]
Pine with smooth Leaves [possibly a form with smooth leaves of *Ananas comosus* (L.) Merr., Pineapple]...

(1) For further details regarding the early history of the strawberry from Chile in Europe, etc., *See* Darrow (G. M.), *The Strawberry, history, breeding and physiology,* 25, etc. (1966) New York.

Hedera terrestris flore carneo [*Glechoma hederacea* L., Ground Ivy, a variety with flesh-coloured flowers].'

Knowlton added a second list, headed, 'Plants I shall be glad of when you can part with them', and among the species he required were the following:

'...Dogs Chops Ficoides [*Oscularia deltoides* (L.) Schwantes, native of South Africa]
Guaver [*Psidium guajava* L., Guava, native of the West Indies]...
Genista montis Labani [probably *Genista libanotica* Boiss., native of Syria and Lebanon]...
Origanum Montis Sepili [*Origanum sipyleum* L., native of Turkey]
Lentiscus [probably *Pistacia lentiscus* L., Mastic Tree, native of the Mediterranean region and Portugal]'

TO RICHARD RICHARDSON, ESQ.

Sep 29, 1750(1)
'Hon.d Sir,
I recd yours & have this day sent a box directed for you; Within as below is the List of Plants...such as you wanted; & [I] will take Care about the books both of Mr Catesby & Edwards &c.
At present I am extremely busy & have not time to say any more being just return'd from the North who am

Sir
Your Obedt hble servant

Tho Knowlton
Nymphaea major flore albo. [*Nymphaea alba* L., White Water-lily, native of the British Isles]
„ „ „ luteo. [*Nuphar lutea* (L.) Smith, Yellow Water-lily, native of the British Isles]...
Cotton. [*Gossypium* L.]
Papaya. [*Carica papaya* L., Papaya]
Azaderach. [*Melia azedarach* L., Bead Tree]
Trachelium umbelliferum. [probably *Trachelium caeruleum* L. (Syn. *Trachelium azureum umbelliferum*), Blue Throatwort, native of the western Mediterranean and Portugal].'

(1) Richardson corres. MS. RAD. TR.C.11. This letter is not in Knowlton's hand.

TO RICHARD RICHARDSON, ESQ.

March ye 8 1750/1[1]
'D^r S^r
I come now to congratulat you one [on] ye hapye alteration you have made & that you may Long Live to enjoy ye fruition of ye same[2]
M^r W^m van Hazen of Leyden has Latly published a Cat: of his garden stoves & green House Plants containing in n^r 850 sorts... there is very near 150 sorts stoves & green house plants in it that we have not in ye English gardens besides a Cat: of ye chedro oranges Lemons & Lemonattas Shadack & Pomplemus to ye n^r of 115 sorts wch being so Large collection I am tempted by it to go over to mak a purchase of al I can... if you have a mind to be a promoter Il do ye best I can for you so, should be glad to hear or have your comands I do not Intend to move that way till ye Later end of aprill of begining of may'

In writing 'chedro' Knowlton presumably misspelt the Italian word cedro, meaning the tree, and fruit, of the citron, and his mention of 'Lemonattas' may be a reference to a small lemon from Italy figured in J. C. Volkamer's *Nürnbergische Hesperides,* 114 (1708) under the name of 'Limoneta da Calabria'. The Shaddock named by Linnaeus in *Systema naturae,* Ed. 12, **2**, 508 (1767) as *Citrus decumanus* and now called *Citrus grandis* (L.) Osbeck, is probably a native of Thailand and Malaysia, and according to Sir Hans Sloane it was first brought to Barbados 'by one Captain Shaddock, Commander of an East-India Ship'.[3] In the East the fruit is known as the Pomelo, supposedly derived from Pompelmoes – the Dutch word for Shaddock. Pompelos or shaddocks are very closely related to the 'Forbidden fruit' or grapefruit. Griffith Hughes in his *Natural history of Barbados,* 125 (1750), referred to the 'Forbidden fruit', and in 1789 Patrick Browne in his *Civil and natural history of Jamaica*, 309, wrote of the 'Forbidden fruit, or smaller Shaddock tree'. Some years later, in 1814 John Lunan in his *Hortus Jamaicensis*, **2**, 172, stated that the 'Forbidden fruit... is considered as only a variety of the Shaddock, but the fruit is much smaller'. In the same work (**2**, 171) the author refers to a variety of shaddock 'known by the name of grape-fruit... not near so large as the shaddock, which received its name from Captain Shaddock...' It is generally assumed that the name grapefruit for the forbidden fruit or smaller shaddock originated in Jamaica, and that it

(1) Richardson corres. MS. RAD. TR.C.11.
(2) Possibly an allusion to Richardson's marriage on 13 February 1750/51 to Dorothy, daughter and heiress of William Smallshaw of Bolton-le-Moors (Bolton) about ten miles north west of Manchester.
(3) *See, A voyage to... Madera, Barbados... and Jamaica,* **1**, 41 (1707).

has been used at least since 1814. In 1830, the grapefruit was separated from the shaddock as a distinct species by James MacFadyen, who gave it the name of *Citrus paradisi* in W. J. Hooker's *Botanical miscellany*, **1**, 304 (1830). As no copy of van Hazen's Catalogue can be located,[(1)] we can only speculate whether 'the forbidden fruit or smaller Shaddock' was included among the names of 115 sorts of Citrus fruits listed in that work.

After telling Richardson that he wished to purchase a number of botanical books that had recently been published Knowlton continued:

'Mr Burry of Salford next week will send me ye Hernandij [*Hernandia sonora* L., Jack-in-a-Box, native of the West Indies and tropical America] a favorit plant of mine I have Longe desired of you... [I] could not Propigate it he [Mr Burry] dos y^m^ by cuttings wch I am shure will do as I have seen before — I hope your stores & green house is in health as all mine are I exspecte ye stripted pineaples plant [possibly a variegated form of *Ananas comosus* (L) Merr.][(2)] in a month time as finely Striped as my aloe wch I Brought first into England. by ye above Cat: I find they have many striped plants that we have not it is ye compleatest [catalogue]...I ever see [consisting of]...a [single] sheet I think I sent you one of his flowers my son Brought over. Mr Edwards has now published his 4th & Last part of Birds wch if you have not will have soon [i.e. the fourth part (1751) of *A natural history of birds,* by George Edwards] I have wort [wrote] Esqr Blackburn word of my designe [to visit Leyden] but as yett have not heard any thing from him

Nither have I askd my Ld yett but if he mends do not dought but hel give Leave by Lett^rs^ from thence he mends fast & is for ye Bath [i.e. to the City of Bath, famous for its medicinal and thermal waters] & Dr Drak [possibly Francis Drake, surgeon and antiquarian] gos with him as Mr Collison [Peter Collinson] tells me in a Lett^r^ Last week from

I am now at Leasure haveing been much engaiged this winter in planting &c & ye wether Improveing temptes me out now & then but in this naiborhood can't mett with any thing agreeable to entertain me the onely gardens in ye northern parts is orford [in Lancashire, belonging to John Blackburn] & Bierly [near Bradford, belonging to Richard Richardson] they Indeed abound with many of ye wonders of ye creation that raised in me a new Life of admireation every vissitt I mak unto y^m^ & such is y^r^ Elegant

(1) William van Hazen gives his address on the title pages of two other catalogues, copies of which are preserved in the British Library, as 'Bloemist op den Marendyk te Leyden'.
(2) *See* Purseglove (J. W.), *Tropical crops. Monocotyledons,* **1**, 77 (1972).

test that few nay very few come in compition with y^{m}... belive me S^{r} y^{r} most obedant Humb^{le} sarv^{t}
Tho Knowlton
P.S.
I hope beloved Botaney will still shear more & more of y^{r} faver & that y^{r} Gardiner continues his Diligence In ye same I hope hel turn out sober & obligeing wch with ye former will render him valuable to so worthy a master & Ldy &c'

TO JOHN FERRET, THE EARL OF BURLINGTON'S LONDON AGENT

22 March 1750/51[1]
'Dear Sir,
I recd your Favour & am very glad to hear my Lord continues in so good a way & hope the Bath will compleat his Recovery [the Earl of Burlington had gone to Bath, in Somerset]

I have long since finished the Ox Close affair all but a Riding or Walk Leading from the small Iron Gate at the end of the Garden at Sleight Hill down the Ox Close thro' the middle of the Thorns into Calf Ing[2] which will have a good appearance from the Garden besides a conveniencey for the leading of the Hay out of the same a nigher way & save the other Closes much from the Damage in Carting: I likewise intend to cut another small way from the upper Court Flat Gate which opens near the Mill houses into the Ox Close which will cross the other walk by which it will be a pleasant Ride down that way to the Closes to shew the Game with much advantage & will cost nothing, for I'll make the Tenants do it who want Thorns to fence wood bridge against the Pales by the Common Roadside. On Tuesday three weeks was the greatest Tempest ever known here by me, in so much that many Trees were overturned & many other broke in two in the Body &c. such was the Havock it made in this Park besides other Damage

As I have a great Desire to see Holland this Spring where is such a Variety of curious Gardens & Plants in them which we have not in this Kingdom, as I find by Catalogues & a Letter from thence prompts me with a strong Desire to be satisfied in the same so humbly beg you'll be so good as write to my Lord to leave for me to make that Excursion which I hope to perform in about three Weeks time if the weather will be so favourable to me for in 8 or 9 Days I can see all the Country & make my Purchases & if a ship be ready to sail, am sure to return in the limited Time & hope you'll

(1) Chatsworth Letter No. 370, in the Devonshire Collections. This letter although signed by Thomas Knowlton, is not in his hand.

(2) Ing: a meadow, or pasture, especially when low-lying near a stream or river, *See The English dialect dictionary... Edited by J. Wright,* **3** (1898).

prevail on my Lord for his favourable Indulgence to me: The Time I would go should be when I have got all my Business into order of my Mind, to go away the end of next Month or Beginning of May & to be home in the same Month & shall think myself greatly obliged to you for that Favour who am Sir

Your obedient hble servant

Tho Knowlton
March 22d 1750/1
P.S. By those little Cuts thro' the Thorns in the Ox Close it will have a Garden Look, as having a little Art mix'd with Nature, so that the One beautifies the other.'

In this letter dated 22 March 1750/51 Knowlton refers to the great storm 'on Tuesday three weeks', i.e. 26 February. In reading of such events it is always interesting to speculate whether the weather on these occasions was merely local or common throughout the country. In the present case it is found that violent storms in various parts of England were reported on 26 and 27 February, so it is clear that the Yorkshire tempest was not merely a local phenomenon.(1) An account of the storms and the damage they caused was published in *The Gentleman's Magazine,* **21**,155 (1751). Here we read of the turbulent weather and the great damage it did t buildings in different counties throughout England, including Nottinghamshire and Yorkshire. Furthermore, in Ireland a hurricane resulted in much havoc in the towns of Limerick and Dublin.

During the many years that Knowlton had been occupied in planting and in making alterations and improvements at Londesborough he needed to employ labourers to carry out his various projects. Unfortunately, these people were not always paid as quickly as they should have been. This is evident from a study of some of the letters written during the years 1741–43 by Burlington's private secretary, Henry Simpson, to the Earl's agent at Londesborough, Ralph Wise. Wise was expected to pay Knowlton the labourers' wages from money received from the East Riding estate. The long periods he kept him waiting for these payments greatly annoyed Burlington. Simpson wrote concerning this state of affairs, as follows:(2)

(1) Information supplied to the present writer by Dr M. J. Ingram, Director of the Climate Research Unit, University of East Anglia, in *epist.* 27.ii.79.
(2) Corres. in Devonshire collection at Chatsworth. Letter Book 42.L.i.

London July ye 11.th 1741
'Sr Mr Wise
Yesterday my Lord gave me a Ɫre wch he reced a little while agoe, from Mr Knowlton wch says that on this day 5 months wages wou'd be due to the labourers... & his Lord-p. comanded me to tell you, immediately to pay so much to Mr Knowlton as will pay those labourers, and for the future to keep ye Same regularly paid...

I am
S.^{r}
Yo.^{rs} &c HS'

London June 18:1743
'S.^{r} m.^{r} Ral:Wise
Last Post but one br.^{t} me a Ɫre from m^{r} Knowlton w.th a months Acc.^{t} ending ye 11: Inst...he says it was the 4.th Month unpaid when upon shewing to My Lord this day his Lordp was extremely angry that the labourers shou'd remain unpaid so many months after his repeated Orders to you to keep them regularly paid and gave me his Orders to write you to pay the same w.thout delay...

I am
S.^{r}
Yo.^{rs} &c HS.'

It was not only the labourers who suffered from Wise's mismanagement. At times Knowlton himself had to wait for money due to him, as is evident from the following letter:

London
Mar: the 27 1742
'Sr Mr Knowlton
I have reced yor Ɫre of the 21 Inst & this day I lay the same before My Lord who Order'd me to write Mr Wise to pay you both for your Labourers & y^{r} wages & board wages... wch I hope will be very soon ...His Lordp Expressed himself very much dissatisfied that Mr Wise had no more regard to his Orders...

I am
S^{r}
Yo.^{rs} &c HS.'

When Knowlton first went to Londesborough in 1726 he was paid, as noted earlier, £12. 4s. 8d. for the 'workmen concerned under him in the garden'.(1) However, by 30 January 1733/34 he was responsible for paying the wages of the labourers employed not only in the garden, but also in the park.(2) Further examples of delay in the payment of labourers' wages are given below:(3)

1.	5 April 1739. A months wages paid Knowlton for labourers, ending 24 February last	£13.16.6
2.	29 May 1739 Labourers imployed by Knowlton, a months wages ending 25 March 1739	£9.12.0
3.	29 March 1739/40 Four months wages paid to Knowlton due to the labourers imployed under him, from 5 November 1739 to 23 February last	£26.2.4
4.	6 November 1740 Paid labourers under Knowlton five months wages, ending 1st Instant.	£44.3.1
5.	12 February 1740/41. Paid several labourers under Knowlton for one months wages, ending 30 November 1740	£7.15.1
6.	25 September 1741 Paid the several labourers imploy'd under Knowlton seven months bills ending 7 September 1741	£59.8.8
7.	27 March 1742/43. Paid the several labourers imploy'd under Knowlton Seven months bills, ending 20 February 1741/42	£26.16.8

No account books are extant which give the wages of individual labourers employed in the park and gardens at Londesborough. In Knowlton's day wages varied according to the distance of the estate from London. For example, in and about the metropolis a labourer could earn 2s. a day, but as far distant as Yorkshire only 1s. or 1s. 2d. a day. Women, who generally did the weeding, received less. In Knowlton's day women were paid about 4½d. or 6d. a day.

Below is given an example of what Knowlton received in wages and board wages. It is found in an abstract of servants' wages and board wages at Londesborough from Lady day 1750 to Lady day 1752:(4)

(1) *See* above, p. 67.
(2) *See* vol. **304** of Londesborough disbursements.
(3) *See The duplicate book for Londesborough begun 8th January 1738/9,* No. 353.
(4) Loose page in vol. **294** in Devonshire collections at Chatsworth.

	£.	s.	d.
Wages	25.	0.	0
Board wages at 7d a week	18.	5.	0
Physick		10.	6
	43.	15.	6
Drawback of Board wages	5.	0.	0

The term 'drawback' indicates the deduction made from annual board wages for the time Lord Burlington and his family, or his family alone, were in residence, that is for those periods when the house staff would not have had to provide meals for themselves but would have been provided for in common by Lord Burlington's cook.

TO RICHARD RICHARDSON, ESQ.

July ye 5 1751(1)
'Dr Sr

Yr Lettr & Box of plants came safe after my departure Yett notwithstand had a copey of it transmited unto me in Holland &c where I was for some time(2) & see Dr Gronovius often(3) who maid enquirey after al his frainds & remember ym over a glasse of wine & desir'd his kind complements to ym al & you amoungest ye Rest mr Tempest ['s] son of Broughton was there at ye same time who has since taken his Degrees(4)

The Gardins of Holland abounding with such numerosos varity of Plants that I belive no Part of ye world do ye Like I could not help purchaseing many of ym but being delay'd in comeing Home again by ye ships at Roterdam [which] pertended sailing from one Day to ye other has accasioned me ye Losse of almost a 100 curious trees shrubs Herbs &c tho I have several now begin to grow a Little & . . . every days shows ym more & more Improveing I Bough[t] Burman Rumfus Hortus amboinenses 6 vol follio but just compleated tho Dr Gronovius tells me he believes he will publish an other & Indeed there is Reason he should do so because he has no Index in those vollums'

(1) Richardson corres. MS. RAD. TR.C.11.

(2) Regarding Knowlton's reason for visiting Holland, *See* above, p. 213.

(3) J. F. Gronovius (1686–1762), a Dutch botanist at Leyden and author of *Flora virginica,* 2 Pt. (1739, 1743) and of *Flora orientalis* (1755).

(4) Henry Tempest (1726–1761), a Yorkshireman, was at the University of Leyden. He took his degree of Doctor of Medicine there 11 June 1751, not long before Knowlton wrote this letter to Richardson. The family were the owners of a property at Broughton, in Yorkshire.

The book mentioned is *Herbarium amboinense* by Georg Eberhard Rumpf, translated by J. Burman from the Dutch into Latin after the author's death. It appeared in six folio volumes in 1741–50, and a Supplement, or *Actuarium*, with an index to the whole, was published by J. Burman in 1755.

Knowlton continued:

'I have several plants for you but none fitt to sent yett as they were a month pack'd up in casses [cases] wch with Losses & charge one [on] customs at Roterdam Hall makes y^{m} very dear to me...

I never was more Pleased with any jorney in all my Life then that Excursion I had not be[en] so much detain'd in my return that gave me much uneaseynes one [on] several accounts... at Last I came to wel in norfolk [i.e. Wells, a town on the Norfolk coast, east of Hunstanton] & so over Land home Leaving al my plants one [on] Board ye shipe I intended to have come in wch did not arrive at Hull till somedays after my Return

I come now to thank you for y^{r} kind parsell of plants & am S^{r}

Y^{r} obediant Humble Sarvant,

Tho Knowlton

PS I am for Lincolonshire to Esqr Burryans at Blybrought to morrow one [on] some new watter works &c'

In 1715 the Blyborough estate was purchased by Zachariah Bourryau, a wealthy owner of property in the West Indies.(1) He died in 1752, the year after Knowlton visited the estate. Green, in his work entitled, *Lincolnshire town and village life,* **2** (1901), under the heading Blyborough, refers to a grand avenue of oaks in the park, apparently the remains of the old drive from the coach road. Moreover, the same authority was told by someone who knew the house well that it was 'an old-world hall, beautifully furnished, and with gardens not to be forgotten'.

Finally, Knowlton closed his letter by telling Richardson:

'Scot ye Gard^{r} at Castle Howard(2) is Turn'd away & they say is gon mad how far trew I do not know but is so said in this part of ye world by many...'

(1) Ross manuscripts vol. **4**, Aslackoe Wapentaka, p. 123. Information supplied by the County Librarian, Central Library, Lincolnshire, in *epist.* 8.xi.78.

(2) Castle Howard, five miles south west of Malton in the North Riding, the property of the Earl of Carlyle.

TO RICHARD RICHARDSON, ESQ.

Octr 1 1751[1]
'Hon^{d} S^{r}

agreeable to y^{r} former request I send you a few of my Holland Plants such I can spare now & shall more in the Spring as they become in health as I have several fine sorts wch I hope to perserve tho I have Lost all ye Hermanias [*Hermannias*] & most of ye Anemonespermus [*Arctotis* L.] with amang other fine sorts yett out of ye maney shall save about half of wch you may be asured of a part in the Spring &c one [on] ye margent [margin] youl see ye charge wch I hope will not displease – by Lett^{r} last week from Mr Collison [Peter Collinson] tells me S^{r} H Slone [Sir Hans Sloane] is well & that the Earl of Northumberland [i.e. Hugh Percy (formerly Smithson), created Earl of Northumberland in 1750] is become a very great Botanist & has a fine collection at Soyon house [Syon, the Earl's estate in Middlesex] & that my Ld Butte is ye most emenant in that way now in England [i.e. John Stuart, third Earl of Bute, an enthusiastic botanist who, after the death in 1751 of Frederick, Prince of Wales, acted as the 'principal manager' to the Princess Dowager in the development of the Royal gardens at Kew].'

After hoping that Richardson and his wife were well Knowlton continued as follows:

'our noble Ld & Ldys with Ld Hartingtons[2] al heare & prity well I have Brought over 3 sorts of ye Larger Kinids of hedgehoges [possibly certain succulent plants of the family *Cactaceae*] with ye South Sea rose very finely striped [*Nerium indicum* Miller or *N. oleander* L., sometimes considered synonymous, native of the Mediterranean region and South Asia. The name South Sea Rose was in use in Knowlton's day, and later for the plant as cultivated in Barbados[3]] . . . nothing grives [grieves] me more then my Leaveing behind me at amsterdam the Pineaple as finly [finely] striped as my alow I Brought 26 years since: but my fear of Losseing it was ye cause otherways the Prise should not have frighted me it was a most charming plant as ever I beheld they told it was brought from Curoso [Curaçao, Lesser Antilles] but ye year before; I see onely one plant & one stole [stool] with 2 smal succers one [on] it, ye Larger I might have had

(1) Richardson corres. MS. RAD. TR.C.11.

(2) The Earl of Burlington's daughter Charlotte Elizabeth married in 1748 William Cavendish, Marquess of Hartington, later fourth Duke of Devonshire.

(3) *See* Hughes (G.), *The natural history of Barbados,* 218 (1750) and Browne (P.), *The civil and natural history of Jamaica,* etc., 181 (1756).

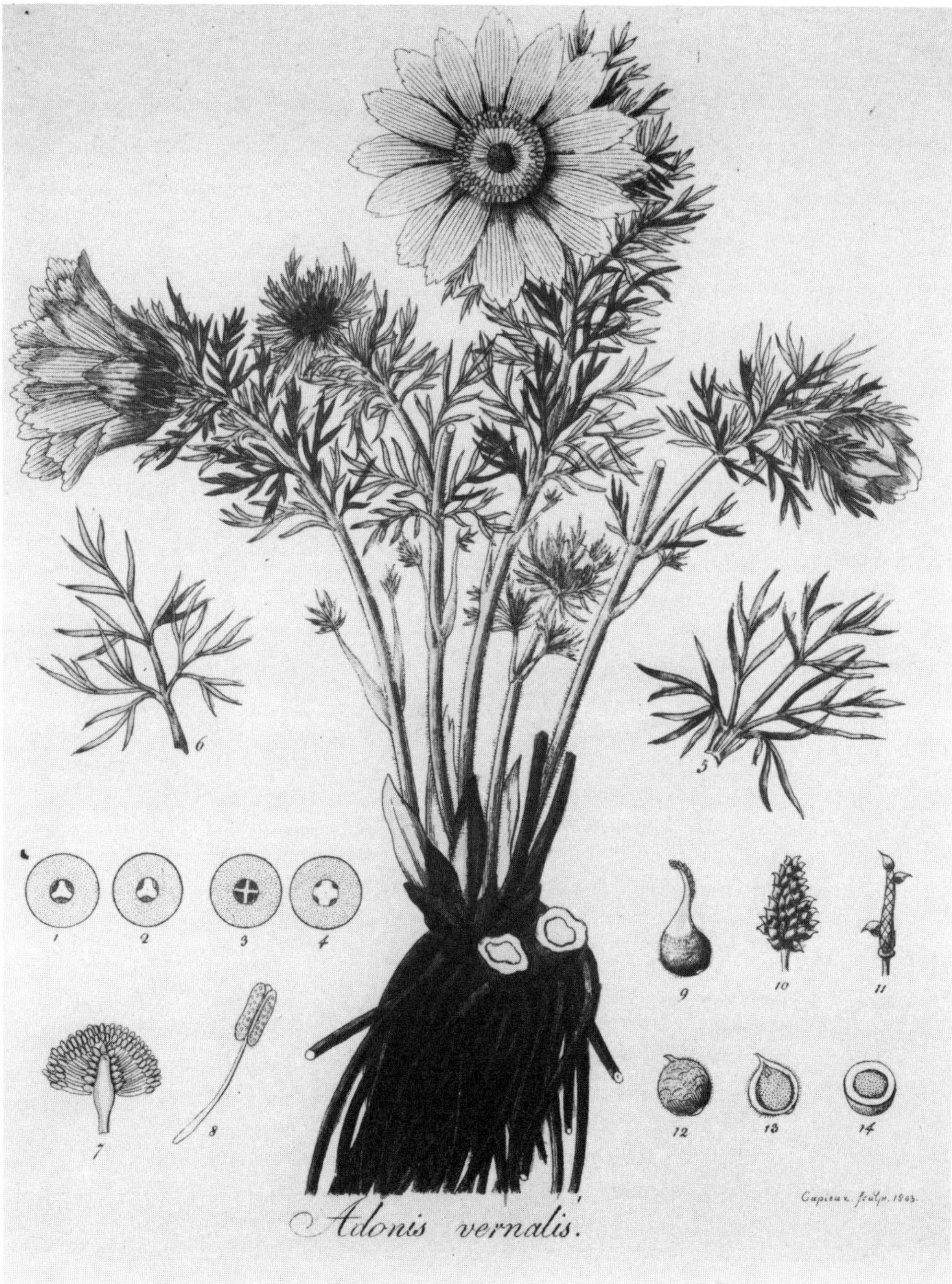

Adonis vernalis L., *a yellow-flowered perennial species, native from central and southern Europe to Siberia, was first introduced to gardens in Britain in the early seventeenth century.*

for 1 — geneey & half neither was it Dear but ye hazed [hazard] was ye cause as I know al strips [stripes] are tender[er] then [than] ye plant of the same kind.

the Box was Sent this Day Down to Weighton [Market Weighton, one and a half miles from Londesborough] to go by ye Hull Carier to York derected for you as usual...'

The species of *Ananas*, Pineapple, coveted by Knowlton that was so finely striped is not known, but it was possibly a variegated form of *Ananas comosus* (L.) Merr.(1) Bradley, in his *General treatise of husbandry and gardening*, **1**, 207 ([1721–]1724) records that some of the plants in the Amsterdam gardens 'were brought thither from that Dutch factories in the East-Indies; but the greatest number of sorts came from Surinam and Curasau... in the West-Indies'. Moreover, the same authority relates(2) that in the garden of Sir Matthew Decker, at Richmond, in Surrey, he had observed some pineapple plants 'very remarkable in their leaves, by being Striped with red'.

Knowlton enclosed in his letter a list of 25 different plants, the total cost of which came to 28 gulden 1 stuiver. The list, dated 'Septr 30th 1751', included the following species:

'Atraphaxis inermis foliis undulatis. H.E. [*Polygonum undulatum* (L.) P. Bergius native of South Africa]
Clutia foliis petiolatis mas [*Clutia pulchella* L., native of South Africa]...
Grewia seu Guidonia [i.e. Guidonia ulmi foliis, flore roseo, now called *Samyda dodecandra* Jacq., native of the West Indies]
Commelina procumbens flore luteo. [*Commelina africana* L., native of South Africa]...
Coma aurea frutesc: foliis glaucis et in xtremitate trifidis H.A II 97(3) [*Chrysocoma tridentata* DC., native of South Africa]
Linaria boetica angusti folia fl: atrorubentes [probably *Linaria aeruginea* (Gouan) Cav., native of southern Spain]
Convolvulus major rectus creticus argenteis [*Convolvulus cneorum* L., native of the west and central Mediterranean region]
Tetragonocarpos afr: frutesc: fol: longis & angustis. H:A II 205(4) [*Tetragonia fruticosa* L., native of South Africa]

(1) *See* Purseglove (J. W.), *Tropical crops. Monocotyledons*, **1**, 77 (1972).
(2) *ibid.* **1**, 209.
(3) i.e. [Commelin (J.)], *Horti medici amstelodamensis rariorum... plantarum... descriptio et icones... Pars altera... auctore C. Commelino*, 97, pl. 49 (1701) Amsterdam.
(4) *ibid.* 205–6, pl. 103.

Clinopodium frutescens foliis linearibus [probably *Pycnanthemum virginianum* (L.) Pers., native of the south-eastern United States of America]

Capnorchis [*Dicentra cucullaria* (L.) Bernh., native of the United States of America]

D. White Throatwort Campanula [probably a double form of *Campanula trachelium* L., native of much of Europe, commonly called Throatwort]

Blue ditto...

Striped Geranium with a fine flower [*Geranium versicolor* L., native of S. Italy, Sicily, Serbia and the Caucasus]...

Adonis hellebori radice buphthalmi flore [*Adonis vernalis* L., Yellow Adonis, native of eastern, central and southern Europe]...

Lotus (S.^{tus} Jacobus) angustifolia flore luteo purpurascente. H:A II 165(1) [*Lotus jacobaeus* L., native of the Cape Verde Islands]

Hyacinthus afr: tuberosus flore caeruleo umbellato. H:A II 133(2) [*Agapanthus africanus* (L.) Hoffsgg., African Blue Lily, native of South Africa.]

Verbenia americana altissima. Par. Bat. 242(3) [*Verbena hastata* L., Blue Vervain, native of the United States of America]...'

The total cost of these plants, namely 28 gulden 1 stuvet, was about £2. 11s. 4d. in English money, as one Silver gulden was equal to 22 English pence. During the next 15 years the value of the gulden remained fairly constant because when Thomas Snelling published in 1766 a work entitled *A view of the coins at this time current throughout Europe,* he recorded that 1 silver gulden of the United Provinces of the Netherlands was worth 22 English pence.

TO RICHARD RICHARDSON, ESQ.

June 2 1752(4)

'Dr S.^{r}

One [on] ye other side youl see ye Bill & List of such Hollanders I have sent for you & in good health & ord^{r} (by ye Hull carrier this day) & hope will come safe to you...

I have Likewise sent you proposalls of m^{r} de Costas designe to Publish a History of fossells I do asure you he is ye most Knowing person I belive now Liveing in that way therefore capable of performing ye same to a great

(1) *ibid.* **2.**

(2) *ibid.* **2.**

(3) i.e. Hermann (P.), *Paradisus batavus,* etc., 242 and pl. (1698) Leyden.

(4) Richardson corres. MS. RAD. TR.C.12.

perfection I have subscribed with m[r] Collison [i.e. Peter Collinson] & shall be glad to hear you will'

Knowlton was here referring to Emanuel Mendas da Costa (1717–91), a noted naturalist who studied conchology and collected fossils. He published various works on fossils and shells, and in 1751 issued Proposals for *A natural history of fossils,* Volume **1**, Part 1 of which (all that was published), appearing in 1757. Knowlton also told Richardson:

'I hope al y[r] noble collection is in good ord[r]... mine are so & al my other Plants pines [*Ananas* cv.] I have good n[rs] with plenty of al sorts of stone frut &c

The Plants I now send are very rair as y[e] Calendula amaranthoides & Indeed al the others w[ch] makes y[m] a great Pleasure to me & glad I am that I have so many in a state of health —

I have been al this spring thinking of waiting on you but a very great Hury of Bussness has pervented me as my Ld is dayly sending his ord[rs] & what I have done for him... is a great deal w[ch] so engaiged me I could not possibly be absent from Home

...from S[r] y[r] obedant Humble servt

Tho Knowlton'

Knowlton enclosed in his letter a list of twenty-one different species of plants, the total cost of which came to 27 gulden (or £2. 9s. 6d. in English money). The list, which is headed 'June 2[d] 1752. Plants sent to Rich[d] Richardson Esqr.', includes the following species:

'Calendula caule frutescens perenni [named by Linnaeus (1763) *Calendula fruticosa* and now called *Osteospermum fruticosum* (L.) Norlindh, native of South Africa. This was presumably the rare plant Knowlton referred to in this letter as 'Calendula amaranthoides'. Philip Miller, in the seventh edition (1759) of his *Gardeners dictionary,* wrote that the plant had been 'of late years introduced into the Dutch gardens from the Cape of Good Hope' and that it had been sent to him 'by Dr. van Royen, Professor of Botany at Leyden'. It is possible that Knowlton obtained a plant from the same source, propagated it by cuttings, and was now sending one of the plants raised by himself to Richardson].

Carthamus arb: Hazen [*Carthamus arborescens* L., native of Spain. Knowlton probably purchased this plant from Willem van Hazen, nurseryman at Leyden]...

Kleinia caule petiolatis truncatis [*Senecio papillaris* (L.) Jacobs, native of South Africa]...
Halleria fl: ruberrimo S. caprifolium [*Halleria lucida* L., African Fly Honeysuckle, native of South Africa]
Barleria spinosa [*Barleria prionitis* L., native of Tropical Africa]...
Marrubium foliis cordatis subrotundis [*Ballota africana* (L.) Benth., native of South Africa]...
Jacobea S. Doria [*Senecio doria* L., native of southern and central Europe]...
Tithymalus Lini folio [*Euphorbia cyparissias* L., Cypress Spurge, native of most of Europe]
Conyza cretica frutescens folio molli candidissimo tomentosa [*Inula candida* (L.) Cass., White Flea-bane, native of Crete]
Ficus nymphaeae folio [*Ficus nympheifolia* Miller, native of tropical America]
Heisteria. Spartium afr: frutescens ericae folio Com̄elin [*Muraltia heisteria* (L.) DC., native of South Africa].'

TO RICHARD RICHARDSON, ESQ.

april ye 24 1753(1)
'D^{r} S^{r}

by y^{r} Gard^{rs} ord^{r} I send you a part of van Hazen plants at ye Lowest charge & hope will come safe as they are well packed & some of y^{m} small therefore would advice a Little Indulgence till Larger as they are fine Kinds hope will give you Pleasure Therein... I have Likewise sent some old plants as he told me you wanted y^{m}

I shall be extreemly glad to heare they are safe al arrived & am in Great hast

S^{r} y^{r} obedant Humble servt

Tho Knowlton
NB ye Box was sent this day to Hull carier...'

On this occasion Knowlton sent Richardson an account listing the price of 22 different species of plants, the bill for 14 of which came to £1. 5s. 0d. The remainder were not priced. Among the various species listed were the following:

'... Gloryosa [probably *Gloriosa superba* L., Malabar Glory Lily, native throughout the Old World tropics]...

(1) Richardson corres. MS. RAD. TR.C.12.

Alkekengii [*Physalis alkekengi* L., Winter Cherry, native of Europe]
Anemonespermus flore Lutteo [probably *Arctotis aspera* L., native of South Africa]
apocynum erectum [*Ichnocarpus frutescens* (L.) R. Br., native of India, Java, etc.]
asparagras [*Asparagus officinalis* L., garden asparagus, native of Europe, including Britain. Regarding the common name of this plant it is interesting to note that John Evelyn in *Acetaria,* 66 (1699), shortened the name to 'Sparagus' from the Latin Asparagus. Later, it was further corrupted to Sparrowgrass]
Coma aurea [*Chrysocoma coma-aurea* L., native of South Africa]...
Caralina Honeysuckle [*Lonicera sempervirens* var. *minor* Aiton, native of the United States of America from New York to North Carolina, commonly called Carolina Trumpet Honeysuckle]
Digetalus accanthoides [*Isoplexis canariensis* (L.) Loudon, native of the Canary Islands]...
Soldonella [*Soldanella alpina* L., native of Europe]
Lotus S[t] Jaccobii [*Lotus jacobaeus* L., native of the Cape Verde Islands]...
Cutting of Euphorbium Trigonum [*Euphorbia antiquorum* L., native of India]
Ipecuana [the roots of the small Brazilian subshrub from which Ipecacuanha or Ipecac is obtained were introduced into Europe in the seventeenth century but the plant was not identified and named until much later. In the *Transactions of the Linnean Society,* **6**, 137–40 (1802), Brotero published a detailed description together with an illustration under the name *Callicocca ipecacuanha.* The plant is now called *Cephaelis ipecacuanha* (Brot.) Tussac. It is not possible to be certain that this was Knowlton's plant. The roots of another plant, native of South America, *Asclepias curassavica* L., commonly called 'Bastard Ipecacuana', were, according to Philip Miller in his *Gardeners dictionary,* sometimes sent to England as Ipecacuanha. Furthermore, there were various other species which on account of their appearance were falsely called Ipecacuanha, and occasionally sold for it]...'

TO WILLIAM STONEHOUSE, gardener to R. Richardson, Esq.(1)

At he head of the first page of this letter is a transcription of a passage in a note received by Knowlton from Emanuel Mendes da Costa. It reads:

(1) Richardson corres. MS. RAD. TR.C.12. Knowlton addressed a later letter, dated 23 February 1758, to 'M[r] W[m] Stonehouse Gard[r] to Esqr Richardson at North Bierly' etc. *See* below p. 241.

'Lister Cochlitanum anglice . . . liber[(1)] mentions several curious figured Fossils he found in the stone quarries near Lawnborough [Londesborough], pray do not be Lazy, but Look for some for me. Small Belemnites of an Amber Colour said Author says are found in Exceeding great plenty in all the Cliffs as you Ascend the Yorkshire and Lincolnshire Wolds, for above 100 Miles in Compass, as at Spiton [Skipton], Lawnsborough etc. and I remember you sent [Sir John] Hill[(2)] some very fine Belemnites from your Neighboorhood.

I shall not Enlarge further at present, because I really think, I have given you a task for some Months; if too much blame your own goodness, and Friendship and not my rudeness.

I beg to have an Answer from you as soon as you conveniently can, for I Long to hear from you . . . I remain with much Esteem

Dear Sir

Your Oblig^{d} humb. Servt

Emanuel Mendes da Costa'

Belemnites, or 'thunderstones', are sharp-pointed, bullet-shaped fossils, consisting of the internal shell of an extinct cephalopod, allied to the cuttlefish. Beneath the transcription from da Costa's letter, Knowlton wrote to Stonehouse as follows:

Jan:27 1754

'Honest W^{m}

The above is a copy of a Lett^{r} from Mr Decosta to me . . . so bege youl use y^{r} Best Endeavirs to procure me as many of ye Impresions of those stones in y^{r} naiborhood & Hallyfax y^{t} you can & . . . of ye fairist sort . . . that I may oblige my above said fraind

I have wort [wrote] a copey of ye same to my son desireing the same of him in his naiboohood[(3)] wch I hope both him & you will comply with – I have wrot to Newcastle & Whitby . . . so hope soon to Recive a collections . . .

I wrote you Long since but fear as I have not yett heard from you y^{t} it was not Recd . . . pray how Dos ye round Leaved Coffey tree Bear . . . Try . . . to rais young plants of ye same or to Lay some of ye Branches that it may be propigated . . . that way . . . ye comon sort will as I have done it

(1) *See* Martin Lister's *Historiae animalium Angliae tres tractatus,* etc. (1678) in which the third tract is entitled 'Cochlearum Angliae et terrestrium et fluviatilium liber', p. 228.

(2) Sir John Hill (1716?–75), the author of many works on botany and other subjects.

(3) From April 1753 until his death in 1714 Charles Knowlton was Rector of Keighley, West Riding.

my self several times(1)... y^{r} wothey master will... give you Leave I hope to make ye collection of fossills for me at the same time you send y^{m} pray send me a clump of ye Pyroala wch I can Keep in the garden... I have such a desire for it'

Philip Miller in his *Gardeners dictionary* (1752) lists four different species of *Pyrola* (Wintergreen) and states that they are:

'...all of them very difficult to cultivate in gardens: for as they grow on very cold hills, and on a mossy moorish soil; so when they are removed to a better soil, and in a warmer situation, they seldom continue long.'

In the closing lines of his letter Knowlton said:

'I hope you Keep ye anagallis well with all others I have sent you Last year... but never Recd any answer from you... wch I wonderd at my kind compliments atend one [on] y^{r} master Ldy &c & desire youl give me a Line to signifie y^{r} Intention of my request &c I am y^{r} obedant
Tho Knowlton
P.S. ye wether is now very savear [severe] by frost & pray how does Esqr Blackburn do for I have not heard from him since June Last... Beasly whom I sent to Henery Englefield has a very fine place to take care of I supose he has wrot to you it is near Reading not much above a mille from it...'

The above remark seems to suggest that Beasly worked for Knowlton before he went to Sir Henry Englefield, the owner of the property near Reading called Whiteknights. It is possible that Knowlton heard that Sir Henry was needing a gardener through someone at Everingham, an estate adjacent to Londesborough. Both Sir Marmaduke Constable of Everingham and Sir Henry Englefield were Catholics.

(1) The writer knows of no Coffee tree with round leaves. Dr Bernard Verdcourt of the Royal Botanic Gardens, Kew, has suggested that it may have been a cultivar or a sport of *Coffea arabica* – the only species of this genus known in England at that time. Knowlton evidently thought the plant he was asking about was a different species from what he referred to as 'the common sort', i.e. *Coffea arabica*.

TO RICHARD RICHARDSON, ESQ.

July 21 1754(1)
'D^r S^r

One [on] the other side youl see a copey of Mr Edward[s's] Lett^r relating to ye apendix of his work wch if he Lives, designs to publish &c & if you have a mind for ye 4 voll^m... please to signifie where you shall have it delived in London... as I shall in a Little time write to him &c'

This is a reference to Mr George Edwards, the artist and naturalist, who published *A natural history of birds* in four parts between 1743 and 1751; he subsequently issued a Supplement, *Gleanings of natural history*, illustrated with plates of birds, fishes, insects, and plants in parts, between 1758 and 1764. Knowlton continued by telling Richardson:

'this morning I have Recd a Box of several Curious fossills from Mr Roundell of Martain... collected by D^r [Robert] Taylor of Skipaton...'

Danson Roundell (1710–70) a Deputy Lieutenant and Justice of the Peace, was the owner of Gledstone House, at Marton, about six miles west of Skipton in West Riding.(2) Dr Robert Taylor may have been an ancestor of the eminent naturalist John William Taylor (b. 1864), second son of John Taylor of Skipton.

Knowlton after completing this letter added a postscript, dated 'July 23', in which he informed his friend:

'this days post brought me a Lett^r from Mr James Scot [of] Turnham green & tells me he has publishd a cutt of ye Sarracen or Side Sadle flower wch grew in his garden [the price in] cullour 2^s plain 1 shilling if you have a mind to have it or any Body in y^r parts will ord^r you on [one] as he desires me to promote ye sail of it he says he exspects ye mignifficant nymphea next Xmas by a shipe from north america if you remember I told you of it he had seeds but not on [one] come up he tells [me] Dr Watson & Mr Collison [Peter Collinson] is well &c'

James Scot, or Scott, seedsman and nurseryman of Turnham Green,

(1) Richardson corres. MS. RAD. TR.C.12.
(2) Information supplied by the Keeper of Geology, Yorkshire Museum, York, in *epist.* 6.xii.78.

near London, had a large illustration made of *Sarracenia purpurea* L., commonly known as the Side-saddle Flower, as it appeared in his garden 'at the Pinery Turnham green the 18 April 1754.'[1] This native of the eastern United States, with pitcher-like leaves serving as fly-traps, continued in bloom in Scott's garden for eleven days.[2] The engraving of this carnivorous plant, after a drawing by Timothy Lewis, was made by the French draughtsman and engraver François Vivares (1709–80), who spent the greater part of his life in England. James Scott was especially renowned for his Pineapples, and he assured would-be purchasers that 'the plants may be convey'd in health & ripe fruit in perfection any journey of sixteen days.'[3]

We must now return to the subject of the box sent to Knowlton by Danson Roundell containing fossils collected by Dr Robert Taylor. The gardener was 'extremely obliged' to receive the specimens which were sent together with notes and short observations. He evidently told his friend Emanuel da Costa about the gift and, subsequently, on 22 August 1754, at Knowlton's suggestion da Costa who had the reputation of being one of the best fossilologists of his day wrote to Robert Taylor about the *Natural history of fossils* that he was preparing.[4] At the same time he enclosed a copy of the Proposals for publishing the work, and asked permission to correspond with the doctor regarding the fossils to be found in his neighbourhood.

Da Costa received a speedy reply to his request and in a letter, dated 25 August[5], Taylor referred to the few specimens that he had sent to Knowlton through the hands of their 'common freind Dawson [Danson] Roundel, esq.' He pointed out that although he had never intensely applied himself to fossil study he had experienced pleasure in finding them as it were by accident on the road, and in some little excursions he had made in this rational amusement. He also told da Costa that if he could, in any respect, be of service to him in his work he must not hesitate to ask him for his help.

Not very long afterwards Taylor sent da Costa some fossils of which he told the naturalist in a letter dated 25 October 1754:[6]

(1) *See* engraved plate, a copy of which is preserved in the British Museum (Natural History).
(2) *ibid.*
(3) Information in James Scott's Trade Card.
(4) *See* Nichols (J.), *Illustrations of the literary history of the eighteenth century,* etc., **4**, 775 (1822).
(5) *ibid.* **4**, 775–76.
(6) *ibid.* **4**, 783.

'I sent you by Mr. Knowlton some picked shells and entrochi, &c. from Clattering Sike.(1) There are still plenty, but not near so much as I remember, they having been carried away in great quantities for grottos, &c. ... You are wrongly informed concerning them: they are not washed from the bottom of a rock, but lie as it were imbedded in the chiver,(2) upon a coarse lime-stone rock, which has been covered by a heathy soil approaching to moss or peat earth, but there is a constant trickling of water betwixt the soil and rock which makes it be called a Sike.'

Da Costa was very grateful to Knowlton for having introduced him to Robert Taylor, and in a letter to him, dated 16 November, da Costa wrote:

'I write this night to my dear friend Mr. Knowlton, and I shall inform him of the great pleasure I enjoy in your correspondence, and of your friendship in having collected my fossils.'(3)

TO WILLIAM STONEHOUSE, gardener to R. Richardson, Esq.

9ber [November] 5 1754(4)
'D^r^ S^r^

it is Long since I see you & did Imagine that you would have been this way before now ... when you are at Leasure shall be glad that you will send y^m^ fossills y^r^ master so kindly gave me ... & hope you will with care send ... y^m^ ... & ... any of those plants you have to spare may come at ye same time; I wishe you would take a Little trouble to Raise ye round Leaved Coffe tree(5) I have now a hernandia [*Hernandia sonora* L., Jack-in-a-box, native of the West Indies] raisd from a cutting at Last from my own tree

I shall send ye old on [one] over to Holland this next Spring with ye magnolia: James Scot gard^r^ Turnham green has published a printe of ye Side Sadle flower [*Sarracenia purpurea* L.] wch Blow'd in his garden, prise collourd 2^s^ he has sent me ... & his B^r^ Henery has published a fine print of

(1) According to *Bartholomew's Survey gazetteer of the British Isles* (1927) there is a Clattering Ford at Bellbank, near Bewcastle, in Cumberland. The area around Bellbank is carboniferous limestone, and there are many sikes in the neighbourhood; perhaps Clattering Sike was one of those. A sike is a small stream of water, a rill or streamlet, especially one flowing through flat or marshy ground.

(2) According to the Oxford English Dictionary 'chiver' is any kind of stone of a slaty or schistose character.

(3) *See* Nichols (J.), *Illustrations of the literary history of the eighteenth century,* etc., **4**, 785 (1822).

(4) Richardson corres. MS. RAD. TR.C.12.

(5) *See* above, p. 227 footnote.

Knowlege & Lab^{r} [Labour] assisted by ye 4 Elements presenting ye gard^{r} with ye fruts of ye hotest Climates Extreemly well done in ye maner of a place bill & is ye neatest thing of ye Kind I have seen he sent me on [one] down.'

The design on this handsome trade card, published in October 1754, was drawn by the noted eighteenth century illustrator Samuel Wale (1721?–1786) and engraved by François Vivares. Henry Scot, or Scott, seedsman and nurseryman, sold fruit trees, flowering shrubs, and greenhouse plants, and like his brother James he specialised in pineapples. The various goods he sold were advertised on his trade card in two columns in English and French respectively. In English we read:[1]

'Pine apples, raised and sold by Henry Scott gardener, at Weybridge, in Surrey. Where persons mey be supplied with ripe fruit during their season... And as the Chertsey coach goes every day from thence to London, so ripe fruit may be easily sent by that conveyance and plants may be safely sent by the Weybridge boats, to London, twice a week...'

It is likely that Henry Scot and Knowlton first became acquainted when the former was gardener to the Earl of Burlington at Chiswick House. The two men also met occasionally at Londesborough.[2]

An excellent botanical book had recently been published in Scotland and Knowlton advised Stonehouse to see a copy. He wrote:

'Pray have you seen Mr James Justice Book cald ye Scots gard^{r} Director prise 5^{s} sold by Mr Hildyard youl Like it for wch reason I recom̄end it to you as I have to several I had a Long Lett^{r} from him a few week since'

The Scots gardiners director by James Justice (1698–1763) was published at Edinburgh in 1754.[3] Justice was the owner of an estate at East Crighton

(1) For a reproduction of this trade card *See* p. 232 and *Gard. Chron.* Series III, **58**, plate facing p. 318 (20 November 1915).

(2) *See* above, p. 176.

(3) For an article on this botanist and horticulturist *See* Minay (R.), *J. Gard. hist. Soc.*, **1**, No. 2, 41–62 (Feb. 1973); **2**, No. 2, 51–74 (Spring 1974); **3**, No. 2, 37–67 (Spring 1975). For the various editions of this work by Justice *See* Henrey (B.), *British botanical and horticultural literature before 1800*, **3**, entries 881–885 inclusive (1975).

The trade card of Henry Scot, seedsman and nurseryman at Weybridge, 'Knowledge & Lab[r] *assisted by y*[e] *4 Elements'. Knowlton described it as 'y*[e] *neatest thing of y*[e] *Kind I have seen'.*

in Midlothian. He became an Advocate in January 1721 and followed his father as Principal Clerk of Sessions in January 1727. Noted as a botanist and horticulturist he devoted much of his spare time and a great deal of money to his garden, where he propagated and cultivated many trees and shrubs, both native and exotic. He grew *Coffea arabica* L., and Pimento trees, *Pimenta dioica* (L.) Merr., and such tender fruits as Bananas, *Musa* cv., Guavas, *Psidium guajava* L., Pomegranates, *Punica granatum* L., Oranges, possibly *Citrus aurantium* L., and Lemons, *Citrus limon* (L.) Burm. f. He was especially interested in growing Pineapples and claimed that they were first brought to fruit in Scotland by himself. He had also a very large collection of Hyacinths, Tulips, Ranunculuses, Anemones, Auriculas, and other florists' flowers, and twice visited Holland in order to study Dutch methods in the culture of such plants. His book contains a number of references to Dutch florists and to their seed and plant catalogues, the latter being sent annually to England from Holland.

In 1739, Justice, being heavily in debt, had to sell Crighton House with its walled garden and range of hothouses, but he continued to practise botanical and horticultural experiments at his house in Edinburgh and at Justicehall, his mansion in Berwickshire.

Knowlton continued by telling Stonehouse:

'Mr Harison [of Manchester] has sent me ye D[ouble] Callindine very Latly wch I hope to see ye flower of in Spring as he say[s] it will come Double again from seeds —'

In this passage Knowlton may have been writing of a double form of *Chelidonium majus* L., Greater Celandine. If he had been referring to *Ranunculus ficaria* L., Lesser Celandine, it is likely that he would have called it by its English name 'Pilewort', as commonly given in eighteenth century publications, including *English botany* by James Sowerby, **9**, 584 (1799). The gardener next asked his friend:

'... have you gott ye Cinamon tree [*Cinnamomum verum* J. S. Presl,[(1)] native of Ceylon] in y^{r} stoves I hope next year to have ye pimenta or alspice [*Pimenta dioica* (L.) Merr., Pimento, Allspice, native of the West Indies and

(1) Previously named *Cinnamomum zeylanicum* Garc. ex Blume, but the oldest, correct, name is *Cinnamomum verum* J. S. Presl. *See* Backer (C. A.) and Bakhuizen van den Brink (R. C.), *Flora of Java,* **3**, 642 (1968).

Central America] &c prey how go they on at Castle Howard... has Teasdale been over to see you yett at Bierly &c'

This is a reference to Robert Teesdale (d. 1773) who succeeded Scot as gardener at Castle Howard. It will be recalled that in July 1751, Knowlton wrote and told Richard Richardson the younger that he had heard that Scot, gardener to the Earl of Carlyle, had been 'turn'd away' on account of his becoming mad. Teesdale's son also worked at Castle Howard and he became a noted Yorkshire botanist. Knowlton closed his letter as follows:

'I bege my most greatfull compliments to y^r worthy master & Ldy to whom I hope youl not forgett... have you heard from Sr Henery Englefield Gard^r they have fine Temples in that garden as any in ye Kingdome & very well worth seeing

I am with al kind respects to you wishing Health & al other ye Desirable Blessings you can Enjoy in this Life &

am y^r fraind

Thos Knowlton

NB I shall at some time send you some Double China aster Read [*Callistephus chinensis* (L.) Nees].'

In 1751, Peter Collinson raised from seed a Paper Tree, *Broussonetia papyrifera* (L.) Vent.(1) This species, native of China and Japan, was described in 1712 by the German physician and naturalist Engelbert Kaempfer in his work entitled *Amoenitatum exoticarum*. From 1683 to 1685 Kaempfer accompanied a Swedish mission to Isphahan (Esfahan in Iran). Subsequently, he joined the fleet of the Dutch East India Company as chief surgeon. He returned to Europe in 1693 and finally settled near his native town of Lemgo, in north-west Germany, as physician to the Count of Lippe. During his travels Kaempfer visited Moscow, Kazan, and Astrakhan. He went to Batavia (near Jakarta) on the island of Java and on the way visited Malabar, Ceylon, and Sumatra; he also went to Siam and from the autumn of 1690 until November 1692 he was in Japan. After Kaempfer's death his collections, including books, botanical specimens, and drawings, were acquired by Sir Hans Sloane. Among his manuscripts was his *History of Japan,* written in High Dutch, the English translation of which, published in 1727, has been referred to earlier as a book Knowlton wished to buy.(2) The Appendix contains a translation of Kaempfer's

(1) *See* below, p. 243.
(2) *See* above, p. 142 referring to: Kaempfer (E.), *The history of Japan... translated... by J. G. Scheuchzer,* etc., 25–26 (1727) Appendix.

description of the Paper Tree *Broussonetia papyrifera.* A transcription of this was sent by Knowlton at the beginning of a letter he addressed to Peter Collinson:(1)

TO PETER COLLINSON

'2 February 1755'

KAADSI.

Papyrus fructu Mori celsæ ; sive Morus sativa foliis urticæ mortuæ, cortice Papyrifera.

The PAPER-TREE.

FROM a ſtrong, branched, lignous root riſes a ſtreight, thick, equal *Trunk*, very much branched out, cover'd with a fat, firm, clammy, cheſnut-colour'd bark, rough without, and ſmooth on the inſide, where it adheres to the wood, which is looſe and brittle, with a large moiſt pith. The branches and twigs are very fat, cover'd with a ſmall down, or wool, green, inclining to a dark purple. They are channel'd till the pith groweth, and quickly decay when broke off. The twigs are irregularly beſet with leaves, at two or three inches diſtance, or more, ſtanding on ſlender, hairy, two inch long footſtalks, of a dark purple caſt, and the bigneſs of a ſtraw. The leaves differ much in ſhape and ſize, being ſometimes divided in three, ſometimes in five ſerrated, narrow, unequally deep and unequally divided lobes, reſembling in ſubſtance, ſhape and ſize the leaves of the *Urtica mortua*, being flat, thin, a little rough, dark green on one ſide, and of a lighter green, inclining to white, on the other. They dry quickly, when broke off, as do all other parts of this tree. Strong ſingle fat nerves (leaving a remarkable hollow on the oppoſite ſide) run from the bottom of the leaves towards the top, and ſend out many tranſverſe ribs, almoſt parallel to one another, which ſend out others ſtill ſmaller, turn'd in towards the edges. In June and July come forth the fruits, (*Tab.* XL. *a.*) *ex alis foliorum*, upon the extremities of the twigs, ſtanding on ſhort footſtalks, round, ſomewhat larger than a pea, ſurrounded with long

(1) Preserved in the Library of the Linnean Society of London (*See* Collinson commonplace books).

purple hairs, compoſed of *Acini*, firſt of a greeniſh colour, which turns to a black purple when ripe: The fruit is full of a ſweetiſh juice. I did not obſerve whether or no there are any *Juli* that come before the fruit. This tree is cultivated for the uſe and improvement of the Paper manufactures on hills and mountains. The young, or two foot long twigs, are cut off and planted in the ground at moderate diſtances, about the tenth month, which ſoon take root, and the upper part, which ſtood out of the ground, quickly drying, they ſend forth many fine young ſhoots, which are fitteſt to be cut for uſe towards the latter end of the year, when they are come to be about a fathom and a half long, and about the thickneſs of an arm of a middle-ſized man. There is alſo a wild ſort of *Kaadſi*, or Paper-tree, growing on deſart and uncultivated mountains, but it is ſcarce, and otherwiſe not very proper for Paper-making, and therefore never uſed.

The above quotation is an excellent example of the lengths to which Knowlton went in communicating detailed botanical information to his correspondents.(1) After giving the above transcription he wrote to Collinson, as follows:

'Dr. Sr. as above you have Kempfers Description of his papyrus... w:ch I take to be yr very plant Mr gorden [James Gordon, nurseryman] had from China dos... & ye other he had from there & if he had maid any Increase yett... as you Live in ye center of the City [you] have much greater opertunity then we in ye country have for wch reeson makes me thus solicitous to you for... Information of y:m'

The gardener continued by writing about three forthcoming publications that he wanted. These were, Patrick Browne's *Civil and natural history of Jamaica,* illustrated with 38 plates made from drawings by the celebrated artist G. D. Ehret, and published in 1756. Of this book he told Collinson: 'I am abundantly thankfull for yr Information of Dr P[atrick] Browns acc:tt of Jamaica I have just now planned to be one of the encouragers of that work'. To this remark Knowlton added that he must have the book. He also wanted, directly they were published, John Ellis's *Essay towards a natural history of the corallines,* issued in 1755, and da Costa's *A natural history of fossils,* issued in 1757.

(1) This quotation is reproduced from the published work.

Knowlton described to Collinson some 'specimens' of a water-lily he called 'nymphea flore albo magno' that had been sent from North America to the nurseryman James Scot, who thought that they were from Pennsylvania. Regarding these aquatic plants Knowlton recounted that they were:

'...that sort with circular Large Leafe 14 or 15 Inches Diameter... ye stalk supporting ye Leafe in ye midle flotting upon ye watter & the flower when Dryd we measured 9 Inches & a half, but ye person as sent y^{m} said he measured some 13 Inches Diameter &c most amazing & far exceeding any thing I have ever yett seen or read on the seed vessells Large & seeds almost as bige as a horse bean &c. now this... so extradinery plant I am surprised has not been sent into England yett as so many have specimens of it make it seem to be prity com̄on thire [North America] — therefore hope youl Lay y^{t} obligation on Mr Battrum [i.e. the collector of plants and seeds John Bartram of Pennsylvania] or some curious person there to send it you over, for of al the plants in ye world that I should covett & Desire as ye being so much superior to al others I had prompted Mr Scot who said he hoped to have it before now[1] & had wrot for it when I was with you'

The only species which fit the above description are *Nymphaea tuberosa* Paine, native of the north-eastern United States[2], and *Nymphaea odorata* Aiton, also native of North America and introduced in 1786 into English gardens.[3] According to the size of the flowers and leaves of the specimen measured by Knowlton it seems most likely that he was writing of *Nymphaea tuberosa*. This species was established as being indigenous in the United States of America in 1865, when Paine published his description of it.[4] There appears to be no record of when it was first introduced into England.

Another piece of news that Knowlton told Collinson was about some water-lilies that grew in the stream of Londesborough. Of these, all native to the British Isles, he wrote:

'I have ye Large white [*Nymphaea alba* L.] ye Large yellow [*Nymphaea lutea* L., now called *Nuphar lutea* (L.) Sm.] & ye Smal yellow flore fimbriato

(1) *See* also above, p. 228.
(2) *See Botanical Magazine*, 6536 (1881) and Conard (H. S.), *The waterlilies, a monograph of the genus* Nymphaea, 189–92 (1905) The Carnegie Institution of Washington.
(3) Aiton (W.), *Hortus kewensis*, etc., **2**, 227 (1789).
(4) *See* Conard (H. S.), *op. cit.*, 179–86.

[Nymphaea lutea minor, flore fimbriato,(1) called by Linnaeus *Menyanthes nymphoides* and now named *Nymphoides peltata* (S. Gmelin) O. Kuntze, the Fringed Water-lily] in the Cannell wch runs thro my Kitchen garden where they do so well that every year I mow up cart Loads of y^{m} & Likewise flower in so elegantly plenty that I never yett see ye Like in any other garden, it may be equald but can't be exceld.'

Knowlton found it difficult to leave the subject of the North American Water-lily. He was willing to pay five guineas for 'a good root of it', and so anxious was he to have it growing at Londesborough that he said, 'I never had a stronger passion for any thing in the whole course of My Life'. His letter was a long one, and conscious of this, he wrote apologetically:

'... pray excuse me for running into such a Length but as I have time now can't avoid it, nature is Intirly ... at rest so that Books ... with ye stoves & green house Plants are ye only amusement of ye season wch now will soon take another turn in faver of vegitation, no snowdrops [*Galanthus nivalis* L.] nor acconits [*Eranthis hyemalis* (L.) Salisb., Winter Aconite] yett apear hear but in stove cyclemuses y^{m} Elice [John Ellis?] so kindly Left with you wch was ye alepicum a fine sort mixt with ye witter [whiter] ones & mexican Lillys [*Amaryllis reginae* L., later called *Hippeastrum reginae* (L.) Herbert(2)] now in blown [bloom] are entertaining with some few others.'

The specific epithet 'alepicum' in the above passage appears to have originated with the publications of Abraham Munting who gives 'aleppicum' as a species of *Cyclamen* in the following works, published at Amsterdam, *Waare oeffening der planten,* 395 (1682) and *Phytographia curiosa,* 30, fig. 145 (1713). In a special copy of John Ray's *Historia plantarum* in the British Museum (Natural History) species not found in Ray were written in the margins, and in volume **2**, 1206 (1688), beneath various species of *Cyclamen* are the following manuscript additions:

'Cyclamen Aleppicum flore purpurea odorato Munt.' and 'Cyclamen Aleppi Munt.'

This copy of Ray's *Historia* was used by Sir Hans Sloane as a partial index

(1) Named thus by Ray (J.), *Synopsis... Editio tertia,* 368 (1724).
(2) *See* An appendix by W. Herbert p. 31 (1821) found at the end of *Bot. Reg.* (1821).

to the collections of *horti sicci* in his herbarium. A specimen of 'Cyclamen Aleppicum flore purpurea odorato' is found in volume 139, on folio 45 of the Duchess of Beaufort's collection, and a specimen of 'Cyclamen Aleppi' is in volume 321 on folio 185 of Hermann Boerhaave's collection. Munting's references to *Cyclamen aleppicum* give no descriptive details and the dates of publication of this author's works have no importance for nomenclatural purposes, as they are earlier than Linnaeus's *Species plantarum* (1753), the starting point for botanical nomenclature. Dr Fischer's name *Cyclamen aleppicum* was published, but without a description, in 1824 in Count J. C. von Hoffmannsegg's *Verzeichniss der Pflanzenkulturen in den Gräfl. Hoffmannseggischen Gärten zu Dresden und Rammenau,* 54, but A. de Candolle in 1844 in his *Prodromus systematis naturalis regni vegetabilis,* **8**, 58, still regarded it as a name of doubtful application. More recently Post in his *Flora of Syria, Palestine and Sinai... Second edition... by J. E. Dinsmore,* etc., **2**, 179 (1933), and O. Schwarz in his paper entitled 'Systematische Monographie der Gattung *Cyclamen* L.' in *Feddes repertorium,* **58**, 259, 261 (1955) have both regarded *Cyclamen aleppicum* as a synonym of *C. persicum* Miller.

To return to Knowlton's letter to Collinson, he told him:

'The cortusa mathioli [*Cortusa matthioli* L., native of south-eastern and central Europe] you had of me I had 20 years ago from my old fraind y[e] old D[r] Richardson of north Bierly he then cald it so: so I... took it to be so.'

It appears from this passage that Knowlton may have made a mistake in the plant he sent Collinson. Collinson had some difficulty in getting Bear's Ear Sanicle, as the plant is commonly called, to become established in his garden. He recorded that he first raised it from seed in 1742, subsequently 'lost it for years, recovered it again from Turin, and [it] flowered anno 1765'.(1) Collinson may not have realized that *Cortusa matthioli* is a woodland plant that favours wet rocks and moist glades.

Knowlton continued:

'... the adianthum floridum [*Cryptogramma crispa* (L.) R. Br., commonly known as Parsley Fern] drys Down every winter but Il take care you have a

(1) *See* Dillwyn (L. W.), *Hortus Collinsonianus. An account of the plants cultivated by... Peter Collinson,* etc., 15 (1843).

specimen & plant too if you please. Mr Harrison of manchester can send enought as it is not above 10 miles from him who has ye Double Celindine(1)

I have Latly had several Lett^rs from M^r [James] justice at Edinburgh he has publishd a Book of gardening thire 8° prise 5 shilling a well wrote Book [i.e. *The Scots gardiners director,* etc. published in 1754 in Edinburgh]... you Know he was one of ye greatest conisours [connoisseurs] in all the north of England but had ye misfortune to over do himself he [is] still a great & passionat Lover of Horticulture Botany Blomest. &c.'(2)

When Knowlton wrote of Justice's 'misfortune to over do himself' he was no doubt referring to his being in debt.

At the end of his letter Knowlton made the following request:

'... when ever you see Mr Ehrrat [i.e. G. D. Ehret, the botanical artist] Mr Elice [i.e. John Ellis, the author and naturalist] pray my kind wishes aloud [to] y^m... as ye same to you... who am y^r most obedant Humble sarvant
Tho Knowlton.'

A postscript reads:

'... pray if any thing new in natural History Botany be so Kind as to lett me Know: is van Royen prodromus published yett I have wrot yesterday to W^m van Hazen [the Leyden nurseryman] to know, & for some seeds, plants, &c., he's the most Knowing in gardening I have yett ever mett with.'

In writing of 'van Royen prodromus' it seems unlikely that Knowlton meant this author's *Florae leydensis prodromus,* as this had appeared in 1740, rather that he had in mind *Otia botanica quibus... definitionibus et observationibus illustratum reddidit Prodromus Florae leydensis A. van Royen,* a work published at Tübingen in 1760 by P. F. Gmelin, Professor of Botany and Chemistry at the University in that city. In this work the writer commented on the system which van Royen, Professor of Botany and Medicine at the University of Leyden, introduced in his *Prodromus.*(3)

(1) Mr Harrison was mentioned by Knowlton earlier, as having sent him 'ye Double Callindine', *See* above, p. 233.

(2) For further details regarding Justice *See* above, p. 233.

(3) The arrangement of the *Florae leydensis prodromus* follows a natural system, the key to which was constructed by Linnaeus and van Royen together. This key is one of the first of its kind and leads to twenty natural classes. Every class is described and discussed. *See* Stafleu (F. A.), *Linnaeus and the Linnaeans,* etc., 159–61, fig. 32, 33 (1971) Utrecht.

TO WILLIAM STONEHOUSE

23 feby 1758(1)
'Honest W^{m}

You have surprised me much that any Lett^{r} directed to me & put into ye post should fail of coming if it a frank(2) & not proper they would have Double charged it & so [it would have] come so I Look on that as an excuse &c or some fault by sending to ye post for I belive in 35 years have I never had any but come if properly directed but you at last have, Il forgive it & say no more

please to Lett^{r} y^{r} master Know on Sunday I had a Lett^{r} from Mr Edward telling ... me he had sent a coppey of his apendix to ye Read Loyon in aldergate street ye 9th of this Instant so shall have it in a few days it is a compleat thing in its kind as it is full of newnesses hel see a real good figure of ye Sithian Lamb.'

This was a reference to George Edward's *Gleanings of natural history,* plate 255 of which contains a figure entitled: 'The Scythian Lamb, a sort of moss so called'.

The fable of the existence of a plant-animal variously entitled 'The Vegetable Lamb of Tartary', 'The Scythian Lamb' and 'The Barometz', is one of the curious myths of the Middle Ages. For some of the many stories of this fable the reader is referred to Henry Lee's book *The Vegetable Lamb of Tartary* (1887) and Alice F. Tryon's publication on the same subject in *Missouri Botanical Garden Bulletin,* **43**, 25–29 (1955). Here we need only say that a fern named *Cibotium barometz* (L.) J. Smith, native of south-east Asia and southern China, is supposed to have given origin to the fable. The stem of this species is covered with long silky tawny-gold hairs which, when old, look like wool. By judicious manipulation the natives of southern China and Formosa (now Taiwan) converted a portion of the stem into a rough resemblance to a lamb, the stem being inverted and supported on the petioles of four fronds cut, leaving only the stubs to serve as legs.

Knowlton continued as follows:

'I have just got ye new natural History of Iceland [i.e. a translation published in 1758 of the Danish by Niels Horrebov] but of al ye paultron

(1) Richardson corres. MS. RAD. TR.C.12.
(2) Stonehouse's letter must have been superscribed by a Member of Parliament.

works[1] I never see any come up to it. I am sick of fulsomeness — I should be extreemly glad to hear that ye Esqr [Richard Richardson] had some spiritt to keep up & Increase his collection of plants & wonder Mr Blackburn did not send by you ye cinamond tree [*Cinnamomum verum* J. S. Presl, Cinnamon Tree, native of Ceylon] when he had ye Kampher so long before [*Cinnamomum camphora* (L.) J. S. Presl, Japanese Camphor Tree, native of China and Japan] – but perhaps hed... Little or Less Inclynation to part with it if he had pray why could you not have brought it with you... I am sorrey you can neither propigate ye coffy[2] or Heinandis [*Hernandia sonora* L., Jack-in-a-Box]... that Jassmine [*Gardenia jasminoides* Ellis, Cape Jasmine] as flowerd at woodford: [in the garden of Richard Warner] Mr Gorden & Mr Collison said was elegant & worth seeing & they told me [John] Miller had figurd it [in Philip Miller's *Figures... of plants*, **2**, pl. 180 (1755–60)] – I took in some of ye n^rs but many was so Il done... besides he [Philip Miller] dos not Imply [employ] a good hand if he had Implyd Ehrat I should never have forsaken him who is ye most Inimatable in Europe now'

Knowlton disliked Philip Miller, gardener at the Chelsea physic garden, who published the work in question, and his condemnation of this publication, which finally appeared in 1760 bound in two volumes, was not justified. The illustrations include a number by Georg Dionysius Ehret whom Knowlton recognized as a most outstanding botanical artist. The figure of *Gardenia jasminoides* is by the German draughtsman and engraver Johann Sebastian Mueller (1715–c. 1790), who did much of his work in this country under the name of John Miller.

Richard Warner (1713?–1775), in whose garden *Gardenia jasminoides* flowered, was the younger son of a banker. He was a good botanist and a keen horticulturist, and the author of a catalogue published in 1771 of the wild plants growing about Woodford in Essex, entitled *Plantae woodfordienses*. He lived at Woodford where he maintained a botanical garden, noted for its rare exotics. Later, further details will be given concerning the manner in which he acquired *Gardenia jasminoides,* or the Cape Jasmine as it was commonly called. Knowlton continued his letter by telling his friend about some other unusual plants:

'... ye new & curiosus plants Introduced into ye Garden since 1753 [when the first edition of Linnaeus's *Species plantarum* appeared] are very great n^rs

(1) The noun poltroon, incorrectly spelt by Knowlton, is here used as an adjective meaning contemptible.

(2) Knowlton was possibly referring to the 'Round-leaved Coffee Tree', *See* above, pp. 227 and 230 and below, p. 246.

but ye most rair is ye periwinkel tree with a purple flower [i.e. *Vinca rosea* L., now called *Catharanthus roseus* (L.) G. Don] & ye adrachne or strawbury tree [*Arbutus andrachne* L., Andrachne, native of south-eastern Europe and Asia Minor] in parkington & Gerard with uncut Leaves a raire tree [i.e. in John Parkinson's *Theatrum botanicum*[1] (1640) and John Gerard's *Herball,* Johnson's edition, 1496 (1633)]: too [two] or 3 sorts new magnolias &c ye paper tree of Kamphora [i.e. Kaempfer's *Amoenitatum exoticarum,* 471 & plate (1712)] &c &c'

Catharanthus roseus is described by J. E. Smith in Rees's *Cyclopaedia,* **37** (1819) [1817][2] under *Vinca rosea,* Madagascar Periwinkle. This erect woody herb, a native of Madagascar, now widely naturalized in the tropics and growing sixty to eighty centimetres in height[3] was, according to Smith, cultivated in England by Thomas Knowlton before 1756,[4] and by the time Rees was publishing the *Cyclopaedia* it had become 'a very popular stove-plant, flowering most part of the year, and recommending itself to general admiration, by the beautiful colour of its ample blossoms', etc.

The Paper Tree, *Broussonetia papyrifera* (L.) Vent., mentioned above was also called the Paper Mulberry Tree. When Peter Collinson introduced it for the first time into England he wrote the following memorandum:[5]

'1751. Raised this year a Mulberry, from seed from China; it proves the Paper Mulberry described by Dr. Kaemfer, in his Amoenitates exoticae,[6] which I have given to the Chelsea garden, to the Duke of Argyle, Lord [i.e. Earl of] Lincoln, etc.'

Knowlton further informed Stonehouse:

'I have Latly some raire sorts [of plants] come down from London & most of y^m^ doe well & I am a goeing to write for a choyse collection for a

(1) *See* p. 1490 where the illustration of this plant is entitled 'Adrachne, The Strawberry tree with smooth leaves'. The plant was usually referred to as 'Andrachne'.

(2) Regarding the date 1817 for this volume *See J. Bot. (London),* **34**, 311 (1896).

(3) *See* Gooding (E. G. B.), Loveless (A. R.) and Proctor (G. R.), *Flora of Barbados,* 328 (1965) and Adams (C. D.), *Flowering plants of Jamaica*, 589–90 (1972) Mona, Jamaica.

(4) Previously, this plant was considered to have been 'first cultivated in this country by Philip Miller in 1757'. *See* Curtis's *Botanical Magazine,* **7**, 248 (1794).

(5) *See* Dillwyn (L. W.), *Hortus Collinsonianus. An account of the plants cultivated by... Peter Collinson,* etc., 7 (1843).

(6) For an English translation of Kaempfer's description *See* above, p. 235.

Catharanthus roseus (L.) *G. Don, a shrubby periwinkle with rose-pink flowers, is native of Madagascar and was brought into cultivation by Knowlton before 1756. It has been widely used medicinally and two of the 70 alkaloids it contains are now employed in cancer chemotherapy.*

gentleman in april next of above 150 raire sorts... I think I told you of y[e] plantain tree being now in frut at Esq[r] Thompson[1] & in Excelent ord[r] but then he has a fine stove 25 feet high al crown glass in front & slope that in sumer it draws with an equality of y[e] hotest climat in East or West Indies so that his pines are beyond any body Else — my compliments wait on y[e] Esq[r] & Ldy to whom I wish health & Long Life & am y[r]

real fraind

Tho Knowlton

At worlyclough Halifax[2]
In yorkshire one m[r] Tho[s] Bolton florist sent M[r] [George] Edwards this time twelve month [ago] a Bird he had not seen before w[ch] he has figured & Described as he is a person of great notice & curiosity... m[r] Edwards says he has had too or three Lett[rs] from him & is much pleased with his correspondance'

Thomas Bolton (d. 1778) was the elder brother of the botanist, artist, and engraver, James Bolton (d. 1799). He collaborated with James in compiling the 'Catalogue of plants growing in the parish of Halifax', published in John Watson's *History and antiquities of the parish of Halifax, in Yorkshire,* 729–64 (1775). In the Department of Botany in the British Museum (Natural History) are preserved examples of water colour drawings of plants by both Thomas and James Bolton.

Knowlton informed Stonehouse that his friend 'M[r] Ellice', presumably the naturalist John Ellis (1705?–1776), had got seed of a *Campanula* from the Canary Islands, namely, *Campanula canariensis,* now called *Canarina canariensis* (L.) Vatke, Canary Bellflower, and that he had stocked the garden with this plant. It had been grown in the Royal garden at Hampton Court as early as 1696 but in 1758, when Knowlton wrote his letter, this species was growing 'in plenty in y[e] gardens'. It is of interest to note that Philip Miller in the seventh edition (1759) of his *Gardeners dictionary* also pointed out that of late years great numbers of the plants had been raised from seed brought from Europe, where it had been many years cultivated, but that the flowers of these new-raised plants 'were not so well coloured as those of the old ones'.

(1) This may have been a member of the family of Thompson of Kirby Hall, Sheriff Hutton, in North Riding, Yorkshire, or of the family of Thompson of Escrick in East Riding.

(2) Warley Clough, about two and a half miles south west of Halifax, West Riding, Yorkshire.

Knowlton concluded his letter with a list of plants numbered 1 to 13 and a note reading:

'7, 8, 12, 13, all grow not far from you so should be glad to know what or wch you can part with... because then I could tell where to gett y^m.'

The list reads:

'1 Althea folio retundiore [*Althaea officinalis* L., Marsh Mallow]
2 arum Zelanicum folio maximo [*Typhonium trilobatum* (L.) Schott, native of Ceylon]
3 coffy with round Leaves [*See* above, p. 227 footnote]
4 Ldy Sliper [*Cypripedium calceolus* L., Lady's Slipper]
5 martinea. [*Martynia annua* L., native of Mexico]
6 marantea [*Maranta arundinacea* L., Arrowroot, native of Brazil]
7 pyrolia rotundifolio [*Pyrola rotundifolia* L., Larger Wintergreen]
8 — S[ive] alsinanthemum [*Trientalis europaea* L., Chickweed Wintergreen]
9 Hernandia [*Hernandia sonora* L., Jack-in-a-Box, native of the West Indies]
10 Jacea Lacinator amsterdam [possibly *Centaurea cineraria* L.,(1) Dusty Miller, native of Italy and Sicily.]
11 Buepleuriodes que simplex [*Phyllis nobla* L., native of the Canary Islands]
12 vitis Idea Roseis minoris foliis [*Vaccinium vitis-idaea* L., Cowberry, Red Whortleberry]
13 Green Leaved maiden haire [*Asplenium viride* Hudson, Green Maidenhair Spleenwort]'

In a postscript the Bierley gardener was told:

'I would not... have you... send me anything till you Lett me know by a List of what you can... the first plant [*Althaea officinalis*] grows plentyfully

(1) Knowlton was possibly referring to Commelin (J.), *Catalogus plantarum horti medici Amstelaedamensis,* 169 (1702) and to the plant listed under the name of 'Jacea Cinerea laciniata flore purpureo', later named by Linnaeus *Centaurea cineraria.* Commelin also listed 'Jacea laciniato Sonchi folio, sive Jacea latifolia purpurea', now called *Centaurea sonchifolia* L., but this plant was not introduced into England until about 1780.

near masden [Moston, a few miles north east of Manchester, in Lancashire(1)] & between it & manchester as I have found [it] & brought it home but have Lost it therefore should be glad of it if you have it...'

TO RICHARD RICHARDSON, ESQ.

March 9 1758(2)
'Most worthy S^r^

as I have just Recd Mr Edwards Book & a dozen of his proposals send you one(3)

youl Like it when you see it as it is full of many curriose subjects wch will give a pleasure to you &c

I hope y^r^ gardens florish & all y^r^ plantations rise that your canells Improve by new additions to y^m^

I Longe much to see y^m^ & you & shall have ye Liberty & time to do so in the Springe wch is now aproching for ye snowdropes acconits Hepaticas and soldenella with ye Chrismasrose all in flower wch foretell ye same

I hope you have M^r^ DaCostas Book of fossills Long since published(4) by wch it appears hes taken Great pains & is esteemed by ye Knowing ones to be a good performance & supose you have Mr Elice [i.e. John Ellis] one [on] ye marin plants or corolin's a very curious one as he is a person of great Knowledge & Industry [i.e. John Ellis's *Essay towards a natural history of the corallines, and other marine productions of the like kind,* published in 1755.]...

My compliments attend y^r^ Ldy & belive S^r^ y^r^ obidant Humble serv^t^
Tho Knowlton'

TO RICHARD RICHARDSON, ESQ.

9ber [November] 13 1758(5)
'D^r^ & worthy S^r^

I belive I am D^r^ [debtor] to you for a Lett^r^ tho can't tell exactly but some time agoe I dined with my fraind Mr Broadly who told me that you & Ldy

(1) Knowlton probably passed through Moston when he went to visit Blackburn at Orford Hall, near Warrington, in south west Lancashire.
(2) Richardson corres. MS. RAD. TR.C.12.
(3) i.e. George Edwards's *Gleanings of natural history* issued in three parts from 1758 until 1764.
(4) i.e. E. M. da Costa's *A natural history of fossils* (1757).
(5) Richardson corres. MS. RAD. TR.C.12. It should be noted that this letter was originally transcribed by Dawson Turner in *Extracts from the... correspondence of Richard Richardson,* etc., (1835) with the date incorrectly given as 'November 13th. 1750'.

was all well & that you enquir'd after me &c. I have maid bold to advise you with [this] hopeing to find you all well as I am now tho' our fraind Dr Drak [Francis Drake] is but very Indiferent...'

Francis Drake, a Fellow of the Royal Society and of the Society of Antiquaries, was the author of *Eboracum: or, the history and antiquities of the City of York,* published in 1736 and dedicated to the Earl of Burlington. After being apprenticed to a surgeon in York he later practised in that city where, in 1727, the corporation appointed him City Surgeon. He devoted much time during his life to historical and antiquarian research and in 1745 he inspected, it will be remembered, ruins near Millington on the Wolds, about two miles from Pocklington, which at the time were thought to be on the site of the ancient Roman town of Dolgovicia.[1]

Knowlton returned in this letter to the subject of the Cape Jasmine, *Gardenia jasminoides,* regarding which he wrote as follows:

'... ye 6 of oct^{r} I had a kind Lett^{r} from Rich^{d} warner Esqr of woodfoot row whos offerd me ye fine Bay Leaved Jasmine [*Gardenia jasminoides*] miller has figured [in his *Figures of...plants* 1755–60]... Last sumer was 12 month [it] flowerd for ye first time in his stove where it was Brought to him by Capt Hutchsion an East Indian man in his return [journey he] cald at ye Cape of good hope for refreshmen & took a ride out into ye country for a days jorney & in his way was most wonderfully surprised by a fine smell & Looking round spied [a] Large Double whitflower wch it came from; ye next day he went with 2 sailors & a Box took it up & planted [it] & Brought [it] to his fraind M^{r} warner who is ye onely one has it... & what is wonderfull ye Dutch Know nothing of it & ye said gentleman has maid me an offer of it & any plant in his Garden duplicated & seeds, wch I might have had Down derectly, but have putt it of[f] till ye spring as being safer &c

Mr Edwards apendix is in much esteem & so is m^{r} DaCostas Book [*A natural history of fossils*] & Likewise Mr Borlace [William Borlase] natural History of Cornwell [1758] if you have y^{m} not, they will give you a very senceable pleasure – I formerly had beged of you & y^{r} gardener to propigate y^{r} round Leaved coffey tree & ye Hernandia [*Hernandia sonora* L.] but fear y^{r} spirit & tast is not so strong as formerly [it] use[d] to be, so that y^{r} gardeners flags I supose likewise for I had very offten desired him to try his skill to propigat y^{m} wch he has not done & I should be extreemly ready to Recd [receive] & Likewise to retaliat ye faver as to ye coffey tree it is not in

(1) *See* above, p. 206.

y[e] Kingdome besides as I know of so if it makes its Exite it will be att an end therefore if it is not done desire youl Lay an Injunction apon him to do it otherways; hel not I am senceable.(1)

the hernandia I had raised a fine plant of a cuting but sent it to van Hazen [nurseryman] at Leyden . . . but . . . from this time to that, I have been without it tho M[r] Warner tell me he exspects [it] this winter over & will send me some of y[m] & china seeds in abundance, Esq[r] Thompson['s] Gardiner Raisd one plant this sumer of it w[ch] I might have, but as but on, I would not(2)

the Prince of Wale[s] is now about preperations for Bulding a stove 300 feet in Lenth for plants & not pines [*Ananas*] & . . . my L[d] Bute has already seatled a correspondance in asia africa america europe & everywhere he can as to be shure my Ld is y[e] most Knowing of any in this Kindome . . . such is his great abilitys there in; & he is y[e] person as has prompted y[e] young prince; & from such what may not be Exspected & next Spring it will rise & grow apacee as all glasse & frams [frames] will be ready:'

The above passage refers to the gardens at Kew House and to George William Frederick, Prince of Wales (1738–1820), who succeeded in 1760 as George III. The Prince's father and mother, Frederick, Prince of Wales, and the Princess Augusta of Saxe-Gotha (d. 1772), had taken great interest in the planning of the grounds of their residence, Kew House, and after Prince Frederick's death in 1751 the gardens at Kew continued to flourish under the auspices of the Prince's consort who received much advice and encouragement from John Stuart, third Earl of Bute (1713–92). Bute assisted in the scientific development of the Royal gardens and furnished them with many rare and exotic plants. The Princess's son, George William Frederick, inherited his mother's interest in Kew, and after his succession as King the gardens rapidly increased in importance as a scientific institution.

In the following passage Knowlton told Richardson of other fine stoves:

'I have Latly just finished 2 stoves with a Little Green house in y[e] midle of y[m] 206 feet Longe for w[m] constable Esq[r] at Burton near Hull with fire walls 170 Long at y[e] each end of y[m], all now I say compleat & finished about 6 week since & is y[e] greatest in it[s] kind of any I know unless M[r]

(1) Regarding the Round-leaved Coffee Tree *See* above, p. 227.

(2) The 'Thompson' here mentioned is almost certainly the same who had a 'plantain tree' in fruit in February 1758, *See* above, p. 245.

Salvill att Croxdeal [i.e. William Salvin of Croxdale Manor] near Derham [Durham] but thire is 3 or 4 times more fire wals for vines &c thire.'

It seems likely that one of these stoves for William Constable was intended for the culture of Pineapple plants, because a bill totalling £22. 2s. 6d. and dated 15 September 1758,(1) lists the following items: 60 pineapple plants at 5s. per plant, 20 at 4s. per plant, and 20 at 2s. per head, i.e. per crown, together with 63 clay pots – 34 at 3d. per pot and 29 at 2d. per pot.

William Constable (1721–91) was the owner of Burton Constable, a large estate in the Holderness area of East Riding. He inherited the property in 1747 and subsequently spent vast sums of money on improving the house and grounds.(2) It is evident from contemporary nurserymen's receipts that he planted the garden with many trees and shrubs, and it seems likely that Knowlton assisted with the selection and planting of the various species purchased. On one occasion, on 5 May 1759, the Fulham nurseryman Christopher Gray (1694–1764), renowned for his collection of American plants, sent William Constable a receipted bill to the value of £14. 17s. 9d. consisting of a long list of plants occupying two columns of foolscap-sized paper.(3) Among the various species listed were the following:

1 magnolia £2.2.0.... 1 moon seed 1/– [*Menispermum canadense* L., native of North America]... 1 Groundsel Tree 1/– [*Baccharis halimifolia* L., native of North America]... 2 single virgins bowers 2/– [*Clematis virginiana* L., native of North America]... 1 Double Spanish broom 7/6 [*Spartium junceum* L., native of the Mediterranean Region], 1 Anona or Pawpaw Tree 5/– [*Asimina triloba* (L.) Dunal, native of North America], 1 Hamamelis 1/– [*Hamamelis virginiana* L., Witch Hazel, native of North America],... 1 Dahoon Holly 4/– [*Ilex cassine* L., Dahoon, Cassina, native of North America], 1 Angilica Tree 2/6 [*Aralia spinosa* L., Angelica-tree, Hercules Club, native of North America]... 1 Casine or Yappon 5/– [*Ilex vomitoria* Aiton, Yaupon, native of North America]... 1 Judas Tree 1/6 [*Cercis siliquastrum* L., native of the Mediterranean Region]... 2 Tulip Trees 5/– [*Liriodendron tulipifera* L., native of North America], 1 Agnus Castus 5/– [*Vitex agnus-castus* L., native of southern Europe], 1 Cork Tree 1/6 [*Quercus suber* L., native of southern Europe],... 1 Liquid amber 1/6 [*Liquidambar styraciflua* L., native of North America]... 1 Azederick or Bead Tree 2/– [*Melia azedarach* L., native probably of India]

(1) Preserved in the Humberside Record Office, Beverley, press mark DDCC Box 49(2).

(2) For further details *See* the *Catalogue to an exhibition at Ferens Art Gallery, Hull,* (February 1970). Introduced by J. Hall on 'William Constable as patron 1721–1791'.

(3) Preserved in the Humberside Record Office, Beverley, press mark DDCC Box 49(2).

Witch Hazel (Hamamelis virginiana L.), *native of eastern North America, was introduced into Britain in 1736. It flowers in autumn, usually before the leaves have fallen, and is now less widely grown than species that flower on the bare twigs.*

It is interesting to note that in a letter of 20 July 1761 John Dunn, William Constable's London agent, sent him a list of American evergreen and deciduous trees and shrubs, the seeds of which could be obtained from Messrs Lee and Kennedy at the Vineyard nursery, Hammersmith, and as a recommendation to deal with these nurserymen, Dunn wrote: 'I find Mr. Knowlton makes use of them for himself.'[(1)]

Constable took great interest in obtaining specimens of plants to dry and mount, his collection finally filling ten volumes that are now preserved in the Library at Burton Constable. The volumes contain specimens of plants gathered in Yorkshire as well as from the garden of Constable's London house. Some are from abroad, for example, '*Anemone nemorosa*' was obtained 'from the park of Marli nigh Paris', and many plants were received from English contributors, including Lord Petre, Sir John Hill, and Thomas Knowlton. One specimen from Knowlton is of Rice, *Oryza sativa* L.,[(2)] and it is interesting to speculate whether he was growing this plant at Londesborough.

Many bills and receipts connected with expenses incurred at Burton Constable are still in existence, and there are a number for work done, and purchases made, by Thomas Knowlton during the years 1758 to 1765.[(3)] One voucher dated February 1759 is given on p. 253, and the entry 'to inspection and directions' is repeated regularly in later accounts. In June 1760 Knowlton was paid for drawings for a menagerie.[(4)]

Croxdale Manor, the property of William Salvin (1723–1800), whose fire walls for vines Knowlton compared with those at Burton Constable, occupied an elevated position near the banks of the river Wear about three miles south of Durham. The pleasure grounds with the adjacent woods and plantations were laid out with great taste.[(5)] The gardener at Croxdale Manor was for many years John Kennedy. In 1776, he was the author of *A treatise upon planting, gardening, and the management of the hot-house* published at York, by which time, however, he was working for Sir Thomas Gascoigne of Parlington, near Barwick, in West Riding.

(1) Letter preserved in the Humberside Record Office, DDCC 145/2.

(2) Knowlton's specimen was identified for the present writer by Dr A. Melderis, sometime Deputy Keeper in the Department of Botany, British Museum (Natural History).

(3) Preserved in the Humberside Record Office.

(4) Press mark DDCC 140(2).

(5) *See* Mackenzie (E.) and Ross (M.), *An historical topographical, and descriptive view of the county Palatine of Durham,* etc., **2**, 439 (1834).

'Wm Constable Esqr D^r^ to Tho Knowlton as follows(1)

		£	s	d
1759				
March 9 & 10	to Inspections & Directions . .	1–	1	—
April 3.4.5.6	to Ditto & seating out the flew in ye stove	2–	2	—
10	to a draining spud	—	4	—
June 7 & 8	to Inspection & directions	1–	1	—
Septr 11.12.13	to y^e^ seating out y^e^ 2^d^ stove flews &c	1–	11	–6
9ber 20.21	to Inspection & derections	1–	1	—
Xber—	to several sketchs & drawings	1–	1	—
Jan: 1759	to a Baskett	—	—	–5
15	to 2 wire Ridles	—	7	—
24	to vines & sorts: 17 trees or plants	—	17	—
		£9=	5=	11

NB please to this Bill to m^r^ church & his rectt. shall be a sufficant Discharge for y^e^ same

Tho Knowlton

Fe^by^ 6 1759 reced Contents in full for
y^e^ Use of m^r^ Thos^s^ Knowlton p^r^
me Dor. Church'

Knowlton ended his letter, as follows:

'...pines [*Ananas comosus* (L.) Merr.]...are become as comon as any fruts so what will be next ye magosten [*Garcinia mangostana* L., Mangosteen, native of Malaysia] & jacks [*Artocarpus heterophyllus* Lam., Jak or Jack, native possibly of the Western Ghats, India(2)] wch are fine fruts yett I doe not hear of any attempt bysides ye Late m^r^ Lacur [i.e. Pieter de La Court] of Leyden...[to grow them]...pray how dose ye Ldy Sliper [*Cypripedium calceolus* L.] florish with you & how dose all the old & curiouse plants...I shall be glad to hear [that they] are in being with you: but yett I dought it for if neither master nor man regard y^m^, of course soon all will be att end only such as can't easey depart.

You see what a vain of Humor I have run into & hope youl Excuse as it is a flowe [flow] of good nature and...hope youl Indulge ye Extravagant sally of ye pen...with Respect & esteem S^r^ y^r^ most obedant Humble serv^t^ *Tho Knowlton*'

(1) Press mark DDCC 49(2).

(2) *See* Jarrett (F. M.), *J. Arnold. Arbor.*, **40**, 336 (October 1959).

It will be remembered that on the death of the Earl of Burlington on 3 December 1753, Charlotte, his only surviving daughter and heiress, brought to her husband William Cavendish, Marquess of Hartington, her father's English estates, including Londesborough.(1) Charlotte died on 24 December of the following year, and on 5 December 1755 Hartington succeeded as fourth Duke of Devonshire, and became the owner of the vast Cavendish estates, including Chatsworth in Derbyshire.

After Burlington's death Thomas Knowlton worked as gardener in the service of the late Earl's son-in-law. We learn of some of the work he did during the early years of the second half of the eighteenth century in a memorandum by Peter Collinson. In this, Collinson referred to a letter he received from Knowlton which was later included in a paper entitled 'Notes relating to botany, collected from the manuscripts of the late Peter Collinson, Esq., F.R.S. . . . by Aylmer Bourke Lambert'. This was read at a meeting of the Royal Society held 18 April 1809, and published in the Society's *Transactions,*(2) where it is recorded:

'By a letter (November 28th, 1762) from Thomas Knowlton, gardener to the [fourth] Duke of Devonshire at his seat of Londesburgh near York, and director of his Grace's new kitchen-garden, stoves, &c., at Chatsworth, I am informed that the Duke of Devonshire is now sowing seventy quarters of acorns, that is 560 bushels; an immense quantity: but this year there was the greatest crop of acorns ever remembered. Besides this vast sowing, some hundred thousands of young seedling oaks are planting out this winter: between forty and fifty mean are employed about this work. In the year 1761, as many oaks were transplanted from the nursery, of two, three, and four years old.'

It is also interesting to note from a surviving account book of the 1750s and 1760s, preserved at Chatsworth, that in 1760 one hundred Laurustinus plants (*Viburnum tinus* L.) were purchased for that estate, possibly to make a hedge but more likely to supply ground cover for game, and that also at this time some 10 000 Mountain Ash trees (*Sorbus aucuparia* L.) were planted.(3)

After the death of the Earl of Burlington in 1753, when Londesborough became the property of the Cavendish family, Knowlton, as stated above,

(1) *See* above, p. 71.
(2) *Trans. Linn. Soc. London,* **10**, 277 (1811).
(3) Information supplied by M. A. Pearman, Deputy Librarian at Chatsworth, in *epist.* 2.iv.81.

remained there as gardener. He was employed in the gardens at Chatsworth by the third, fourth, and fifth Dukes of Devonshire. This is recorded by Knowlton's grandson, Thomas Knowlton the younger, in a letter to the sixth Duke, when he wrote,(1) 'My grandfather served your... father, grandfather, and great grandfather.'

TO WILLIAM STONEHOUSE

feb^{y} ye 27th 1763 Londsbro'(2)
'Dr Will

After our long silence I again write to address you with how do you do and your good master and Lady, pray have you the Acetosa rotundifolio Eboracensis [*Oxyria digyna* (L.) Hill, Mountain Sorrel], which grows wild on Longsledale near Buckborrowell [north of Kendel, Westmorland] Ray Syn[opsis, 143 (1724)]! & Hudsons flora anglica [156 (1762)]. A plant I have had in this Garden this 30 years and then from your Masters Father [i.e. Dr Richard Richardson], but now have lost it, and should be glad to restore it again if, that you are so happy as to have it preserv'd for an old acquaintance... I likewise could be glad to know if... you have any collection of the Ficoides [*Mesembryanthemum*] as I want the Micans flore amantio of Dellenious [*Hortus elthamensis,* **2**, pl. 215, fig. 282, i.e. *Mesembryanthemum micans* L., now called *Drosanthemum micans* (L.) Schwantes, native of South Africa] & Hispidium repens flore striato [i.e. *Mesembryanthemum hispidum* L., now called *Drosanthemum hispidum* (L.) Schwantes, native of S.W. Africa], if you should have them, wch I have but lately lost shall be glad to retaliate the favour, Esqr Blackburn has not them now... I rec^{d} several fine sorts of Plants from him... was the later end of summer with him at Orford Hall and see his Collection... [he] has... lately rebuilt his stoves in a more modern way and his plants all in good order,

I presume you know y^{t} Mr Edwards is now about publishing another part of Natural History of Birds [i.e. the third part (1764) of *Gleanings of natural history*]... & no doubt but you have heard of the Welch Society is upon such another subject, as Thos^{s} Pennant is the chief whom perhaps you may have heard of by y^{r} neighb^{r} Thos Bolton near Halifax, he has been long a collector of natural curiositys wch I hope your leisure is sometimes regaled with'

Thomas Pennant (1726–98), traveller, naturalist, and an author of note,

(1) Letter, dated 23 January 1820, preserved among the archives in the Devonshire collections at Chatsworth, No. 41.8.

(2) Richardson corres. MS. RAD. TR.C.12. This letter is not in Knowlton's hand.

was a native of Whitford (or Whiteford) near Holywell, Flintshire. In 1766 appeared in London his *British Zoology*, on quadrupeds and birds, published by the Honourable Society of Cymmrodorion, a society founded in 1751 for the encouragement of the arts, literature and science of Wales and research thereon. Knowlton also told Stonehouse:

'I am as Zealous as ever in the Botanick way and have dayly new amusements from that subject from the infinite varietys which North America now affords and furnishes us with, as the noble tribe of Magnolia's y^{e}, altissimas [*Magnolia grandiflora* L., Bull Bay][1] with ye umbrella tree [*Magnolia tripetala* L., the leaves are crowded in an umbrella-like circle at the summits of the flowering branches] and the new mountain sorts [*Magnolia acuminata* L., Cucumber tree, Mountain Magnolia], with the little [*Magnolia virginiana* L., Sweet Bay, Swamp Bay, Small or Laurel Magnolia] all which flower yearly in those Gardens which are possess'd with them, particularly my friend Mr Peter Collison at Mill Hill near London wch I do assure you flowers profusely there.'

Peter Collinson himself recorded, that he grew *Magnolia grandiflora* from seed and after 20 years it flowered in 1760. *Magnolia tripetala* was sent to him from South Carolina in 1753 and he had 'it planted in Mr. J. Gordon's garden [at Mile End] for him to encrease it'. On 24 May 1760 this tree had the distinction of being the first of the species to flower in England. Collinson wrote of his *Magnolia acuminata*:

'May 20, 1762, the Mountain Magnolia flowered in my garden, which I raised from seed about twenty years ago, being the largest and highest tree in England, and is yet in very few gardens.'

Magnolia virginiana L. (Syn. *Magnolia glauca* L.) was cultivated in England as early as 1688 by Henry Compton, Bishop of London, and regarding this species Collinson recorded:

'Magnolia minor, or Swamp Bay; raised by Lord Petre, and flowered in 1742.'[2]

(1) Catesby called this tree, 'Magnolia altissima, flore ingenti candido' *See The flora of Carolina,* etc., **2**, 611, pl. 61 (1738).

(2) *See* Dillwyn (L. W.), *Hortus Collinsonianus. An account of the plants cultivated by... Peter Collinson,* etc., 32–33 (1843).

To return to Knowlton's letter, he wrote to his friend:

'shoud be glad to hear that your Gardens flourishes, with all that they contain likewise the plants that was scattered in y^{r} Rocks as they were likely places to grow upon and to be so uncommon a natural ornament to ye same & pray have you lai'd open the older Garden yet for tho it was my favourite for a Botanick one and once was for its bigness the most compleet one in the World for rare & curious sorts of the vegetable tribe now had you ye like in one of y^{r} vallies in y^{r} retirements in your walks how pleasing & entertaining would it be nay an agreeable surprise likewise I assure you such a mixture with the other superb works could not be amiss & perhaps you may see at Chatsworth somewhat of this kind in a little time. In order to compleat ye noble amusements of that place I think [you] should attend in all such places, as an Instance at Kew ye presant [home of the] Princess of Wales where there is one of the best Collections in the Kingdom if not in the World &c so we cant say she wants Taste, – As I have run on hope y^{t} you'll both forgive me as well as let me have a line from you . . . y^{r} real friend

Tho. Knowlton

NB. at Gardelupe [Guadeloupe, Lesser Antilles, West Indies] is plenty of Cinamon trees [*Cinnamomum verum* J. S. Presl] as I have seen one from thence last sum̄er at James Gordon Mile End London . . .[1]

P.S. I hope y^{r} worthy masters Mother & Brothers are well to whom I wish them health &c My friend Dr Drake is but indeferent he's very often at Bramham Park Ld Bingleys but at York just now'

Francis Drake had already been suffering indifferent health when Knowlton addressed a letter to Richard Richardson on 13 November 1758.[2] Drake often stayed at Bramham Park. The fine mansion built at the time of Queen Anne belonged to George Fox, afterwards Fox-Lane (d. 1773) who was created in 1762 Baron Bingley. His estate was about three miles west of Tadcaster, in West Riding.

TO RICHARD RICHARDSON, ESQ.

May day [1763?][3]

'Most Hon^{d} & worthy S^{r}

I had wrot to you Long since &c. & now haveing this opertunity bege

(1) The Cinnamon tree, although grown in the West Indies, is a native of Ceylon, South India, and Malaya.

(2) *See* above, p. 248.

(3) Richardson corres. MS. RAD. TR.C.12. The letter was written about the time of publication of Linnaeus's *Species plantarum . . . Editio secunda,* **1** (1762), **2** (1763). *See* below, p. 260.

Leave to Adresse you & Ldy with how doe you doe &c & can't help Informing you of ye many wonders in gardening & to what a pitch of perfection it is arrived too [to] that many of ye noblemen are become as Knowing in ye raire plants both in stoves greenhouse & gardens as to sarprise & please me at ye same time as I have been a wittness of many viz as my Ld Bute Straford Lincolon Coventry Buchanon & many others of Late & In Short it would be Endlesse to tell you what a sperrit there is in gardining near ye Great city of London & thereunto ajacent'

Who were the five peers mentioned by Knowlton and where did they own estates? John Stuart, third Earl of Bute, has already been referred to with regard to the part he played, after the death in 1751 of Frederick, Prince of Wales, in giving the Dowager Princess Augusta advice and assistance in running the gardens of Kew House.(1) Bute devoted much of the earlier part of his life on his estates in Buteshire tree-planting and studying botany. He also made improvements in the gardens at Caenwood House (now Kenwood House) near London, before he sold the property in 1754. In 1763 he purchased the estate of Luton Hoo, Bedfordshire, where he made a botanic garden.

William Wentworth, second Earl of Strafford (1722–91) inherited in 1739 Wentworth Castle, at Stainborough, about three miles south west of Barnsley in West Riding. After his accession to this estate the Earl made many alterations. He did away with the old formal gardens and proceeded to landscape the grounds. A shrubbery was planted near the house, plantations and woods were laid out, winding walks and graceful slopes were introduced, water was made to play a prominent part in enhancing the beauty of the landscape, and temples were placed to command views over the park and countryside. Kenneth Lemmon in writing about this estate has quoted from the local Barnsley historian, Joseph Wilkinson, who recorded that the alterations to the grounds at Stainborough were carried out under the superintendence of the eminent eighteenth century landscape architect, 'Capability' Brown.(2)

Henry Fiennes Clinton, afterwards Pelham-Clinton, ninth Earl of Lincoln (1720–94), inherited the Earldom and the estate of Oatlands, Weybridge, Surrey, in 1730. In 1768 he became the second Duke of Newcastle-under-Lyme. As a young man Lincoln began re-designing

(1) *See* above, p. 219.

(2) *See* Lemmon (K.), Wentworth Castle: a forgotten landscape, *J. Gard. hist. Soc.* **3**, No. 3, 50–57 (Summer 1975).

the gardens at Oatlands from a formal to a more natural style.(1) It was planned to have large pleasure grounds near the house, surrounded by a park from which fine views were obtained across the valley of the Thames. From a terrace could be seen the lake, three-quarters of a mile in length and outstanding for its natural appearance. In the distance was the countryside. Many trees and shrubs were planted on this property and the number purchased increased greatly during the fifth decade of the century. This is evident from the fact that the amount spent on trees rose from £11. 1s. 0d. during the period 1745–55, to £864. 9s. 0d. in 1758–59. Gardeners' and nurserymen's bills issued from 1765 to 1775 inclusive provide further evidence of the extensive amount of planting that was carried out at Oatlands. During those ten years a total of £499. 13s. 4d. was spent on the purchase of seeds. Lincoln made several annual payments to Peter Collinson's scheme by which the Pennsylvanian botanist, John Bartram, was paid to collect and send to England boxes of seeds of North American plants. The collector received five guineas for each box. These were distributed to the subscribers by Collinson. The scheme continued annually from 1735 until 1760.(2) Among friends from whom the Earl of Lincoln received gifts of plants was Peter Collinson who, it will be remembered, gave him a Paper Mulberry Tree (*Broussonetia papyrifera* (L.) Vent). The Earl of Burlington designed a villa for Oatlands, but Symes(3) has pointed out that there is doubt whether it was ever built.

George William Coventry, sixth Earl of Coventry (1722–1809), succeeded to the title and to the estate of Croombe d'Abitot, Worcestershire in 1738. Croombe d'Abitot lay near Pershore in low-lying marshy ground. With the assistance of 'Capability' Brown Lord Coventry drained the land and subsequently planted it with 'all the useful and ornamental trees and shrubs at that time to be procured in the nursery gardens.'(4)

It is difficult to be sure to whom Knowlton was referring when he wrote of Buchanan. It may have been to William Graham, second Duke of Montrose (1712–90), who also had the title of Marquess of Graham and Buchanan, and whose estate at Buchanan, Stirlingshire, was situated between Drymen and the shores of Loch Lomond. The Duke did a large amount of tree-planting at Buchanan.

(1) For an illustrated Article on Oatlands Park in the eighteenth century *See* Symes (M.), in *Gard. Hist.,* **9**, No. 2., p. 136–56 (Autumn 1981).

(2) *See* Collinson. Autograph MS in British Museum (Natural History) entitled, 'An account of the... introduction of American seeds into Great Britain'.

(3) *Op. cit.,* p. 138.

(4) *See* Loudon (J. C.), *Arboretum et fruticetum,* etc., **1**, 73 (1838) and Dean (W.), *An historical and descriptive account of Croome d'Abitot,* etc., 37–38 (1824).

In an earlier letter, dated 13 November 1758,[1] Knowlton had written to Richardson about the Cape Jasmine, *Gardenia jasminoides,* which had been brought to Richard Warner by Captain Hutchinson from the Cape of Good Hope. In the present letter the gardener had more to tell Richardson about this beautiful plant. He said:

'I Lately was Invicted to dine with Rich^{d} warner Esqr at woodford row & spent a whole day with him see his gloweryous [glorious] Jasmine & Captain Hutchinson who Brought it over to him, & p^{d} him y^{e} adorciation ... dew for so great a jewel wch if it had been in Grays or Gordons hands [i.e. in the hands of the nurserymen Christopher Gray of Fulham or James Gordon of Mile End] they might have maide 1000$^{££}$ [£1000] of it in y^{e} year 1759 to my certain Knowledge for I bid 5 geneiys [guineas] for a smal plant & so did my Ld coventry at y^{e} same time if you have not seen y^{e} acc^{tt} of it youl be surprised as what I have seen is one flower a pure white & 4 Inches diameter Double & Elegantly fragrant by wch ye Captain found it... & [it] was unknown to all in Europe... Linnaeus has just now published a new edition with Large additions of his Specaes plantarum and [it] is come over but about 3 weeks in wch is m^{r} warners Jasmine & all ye new plants could be known hear as ye descriptions have been sent to him by my fraind Mr Elice [John Ellis] who is as Indefatageble as any man I Know [Vol. **1** of the second edition of the *Species plantarum,* was published at Stockholm in 1762, and Vol. **2** in 1763]...'

Warner's 'Jasmine' was named *Gardenia florida* by Linnaeus in the above mentioned second edition of the *Species plantarum,* **1**, 305 (1762), but Ellis had already described and published the same plant earlier, in 1761, under the name *Gardenia jasminoides,*[2] which name, therefore, takes priority according to the *International code of botanical nomenclature.* Although the plant was brought from the Cape of Good Hope by Captain Hutchinson, it was not a native of South Africa but of China and Japan.

John Ellis frequently corresponded with Linnaeus, especially with regard to methods of conveying plants from distant countries. He published several pamphlets on the subject of preserving plants and seeds during long voyages. His best known work, however, in his *Essay towards a natural history of the Corallines* (1755).

(1) *See* above, p. 248.

(2) *See Phil. Trans.,* **51**, Pt. 2, p. 935, and plate entitled: 'Gardenia with a double flower' (1761).

Knowlton ended his letter by telling Richardson:

'M^{r} [George] Edwards showd me 20 new sorts of Birds taken in a franch shipe designed for Mad^{m} de pampadorea [i.e. La marquise de Pompadour] & Brought from ye Spanish main all new & of wonderfull collowerings such I had not ever seen before... I am... y^{r} most... Humble serv^{t}
Tho Knowlton'

TO JOHN GRIMSTON

28 May 1763(1)
'Hon.^{d} S^{r}
Since I Recd y^{r} faver I have M^{r} [George] Edwards Books(2) & sent Esqr Constable his [i.e. William Constable of Burton Constable, East Riding] & I hope y^{t} you have y^{rs} all compleat & y^{t} they give you a pleasure & Great satisfaction... I have not very Latly heard from Esq Pennant [i.e. the naturalist and traveller Thomas Pennant]... y^{r} most obedant
Humble servant
Tho knowlton'

Knowlton numbered among his many correspondents John Grimston (1725–80), a member of a family long associated with Grimston Garth, about four miles south of Aldbrough in East Riding, in the peninsula of Holderness. His father Thomas Grimston (1702–51) inherited the Kilnwick estates, about seven miles north of Beverley, from a relative, Admiral Medley, in 1747. John Grimston was 'of a retiring nature, studious and cultured, a great bibliophile, and a collector of scientific and curious objects.'(3) In later letters, written between 1775 to 1779, Knowlton told Grimston of other publications he thought he would be interested in. After his father's death John Grimston became the owner of the family estate of Kilnwick, and from the beginning of the seventh decade of the eighteenth century extensive planting was carried on there. Plants were purchased from various nurserymen, including Telfords of York, and in 1778 a French Catalogue was obtained listing Apricots, Peaches, Pears, Plums, and Vines.(4)

(1) Grimston correspondence in the Humberside County Record Office DDGR 42/13.
(2) Possibly George Edwards's *A natural history of birds,* 4 vol. (1743–51). and Supplement — *Gleanings of natural history,* 3 vol. (1758–64).
(3) Ingram (M. E.), *Leaves from a family tree being the correspondence of an East Riding family,* etc., 54 [1951].
(4) Harvey (J.), *Early gardening catalogues,* etc., 41 (1972).

TO EMANUEL MENDES DA COSTA

July 19, 1765(1)
'Oh my good Sir,

As I have this opportunity, would not let it slip, to tell you that I received your most valuable present, which I find is abounding with several plates, and that he has copied most of Rumphus on shells in every plate, and well copied they are, and that without taking any notice from whence so taken. I hope ere long I shall have the pleasure of making some returns.'

The 'he' who 'copied most of Rumphus on shells' may be identified with James Petiver who practised as an apothecary at the White Cross in Aldersgate Street, London. He was a keen naturalist and was the author of many works on the subject, including *Aquatilium animalium Amboinae, &c. icones & nomina* (1713). This was later included, together with a Supplement, in the first volume (1764) of a publication consisting of a collection of the author's works published under the covering title of *Jacobi Petiveri opera, historiam naturalem spectantia; or, Gazophylacium.*

George Eberhard Rumpf (1627–1702) better known under the Latin form of his surname, Rumphius, was the naturalist referred to as 'Rumphus'. He spent the greater part of his life on the island of Amboina in the East Indies in the service of the Dutch East India Company, and during this time, through study and research, became unequalled in his knowledge of Moluccan natural history.(2) He did not live to see his two great works on the subject in print. They are his *Herbarium amboinense* (1741–55) and *Amboinsche rariteitkamer* (1705). The latter is written in Dutch and consists of accurate and detailed observations on shells, crabs, and other invertebrates, etc., together with sixty plates. A study of these plates together with the twenty contained in Petiver's *Aquatilium animalium amboinae, &c. icones* reveals that many of the figures in Petiver's work have been copied from Rumphius's *Amboinsche rariteitkamer,* but with no word of acknowledgement. However, it is interesting to note that the Supplement with two additional plates (XXI–XXII) to Petiver's *Aquatilium animalium amboinae, &c. icones* contains a list of the figures depicted on these two plates with details from where in Rumphius's work they were copied.

The copy of Rumphius's *Amboinsche rariteitkamer* in the British Museum (Natural History) belonged to James Petiver and was used by him in the

(1) This letter by Knowlton has been copied from Nichols (J.), *Illustrations of the literary history of the eighteenth century,* etc., **4**, 749 (1822).

(2) Wit (H. C. D. de), *Rumphius memorial volume edited by H. C. D. de Wit* (1959).

preparation of the *Gazophylacium*. It contains some of his manuscript notes.

In the *Catalogue of the botanical and miscellaneous library* of Knowlton's grandson which was sold by auction in 1846,[1] a note under the entry for the two folio volumes (1764) of Petiver's *Gazophylacium* reads, 'Presented to Tho. Knowlton by Emanuel da Costa, June 1765'. This is probably the gift Knowlton was referring to at the beginning of his letter. To show his gratitude the gardener wished to send his friend something in return, but in so doing seems, as Nichols points out, to have forgotten that da Costa was a Jew. He wrote:

'It is said by most people that Yorkshire hams are very much admired, and if you should think so, will send you some up, if that you will please to let me know whether they will be agreeable or no.'

Knowlton continued his letter by giving his opinion of the Cambridge Botanic Garden. It was not complimentary.

'I was at Cambridge, but was greatly disappointed in the design of the Physic-garden, which is laid out so preposterously, that instead of pleasing, gives a disgust; besides, it has little in it of proper furniture, plenty of all the common sorts, designed more to please the ignorant than the curious, for young Miller [i.e. Charles Miller, Curator of Cambridge Botanic Garden 1762–70] is not often to be seen in the garden, as I so understand, having a salary for walking about; but of all the gardens I ever saw, it is the worst furnished. I will say no more...'

The Cambridge Botanic Garden was founded in 1762 through the generosity of Richard Walker (1679–1764), Vice-Master of Trinity College, Cambridge. It was situated in the centre of the city of Cambridge, in the parish of St Edward's, the site being chosen by Walker with the assistance of Philip Miller of Chelsea. Thomas Martyn (1735–1825) who had succeeded his father John Martyn as Professor of Botany at Cambridge in 1762, was appointed the first Reader on Plants in the Garden, and Charles Miller (1739–1817), Philip Miller's younger son, the first Curator. Knowlton's strictures may have been partly due to the fact that this garden was explicity laid out according to the Linnaean system, Thomas Martyn being by that time completely converted to Linnaean

(1) Copy in British Museum (Natural History).

botany. Knowlton, unless his opinion had changed since writing to Dr Richardson, on 31 October 1736, was critical of Linnaeus's so-called 'sexual system' of classification.(1) Furthermore, the Cambridge garden, fairly quickly after it was founded, faced difficulties which were partly financial, and its reputation was not very high until James Donn (1758–1813) became Curator in 1794. When Donn took up his appointment he found the garden poorly stocked. Knowlton, always enthusiastic about rare and exotic plants, was doubtless very disappointed to find the Cambridge garden 'designed more to please the ignorant than the curious' with all the common sorts of plants.

TO PETER COLLINSON

July ye 13 1767(2)
'oh my Good S^r

how much did you rejoyce & please me by y^r Indulgence of a long & Kind obligeing Lett^r wch I doe asure you I have read over & over with Great satisfaction, & am Glad y^t you have maid a reservation of a roome in Grace Church street [London] whare now & then I shall trouble you: I likewise rejoyce y^t you & son are well as I thank God is mine & all as belongs [to] me, my son has not much Inclynation to Botany tho Knows some plants well & has Latly sent me 2 sorts of heath y^t he found one [on] ye moore with ... milk white flowers wch I am realy fond of ... & he Likewise sent or Brought me ye Bunch Bury y^t is ye rubus saxatilis L. [commonly called Stone Bramble, and in Yorkshire and Cumberland, Bunchberry(3)] wch I had never seen before (tho' you have) with several other moreish [i.e. moorish] plants so y^t he is not altogether Ignorant.'

At the time this letter was written Knowlton's son, Charles, was Rector of Keighley in the West Riding, an appointment he held until his death in 1814. He was born in 1727, soon after his father became gardener at Londesborough. After leaving Coxwold Free School, Charles went to Cambridge, being admitted at the age of sixteen as a sizer at St John's College. He took his Bachelor of Arts degree in 1747–48, proceeding to Master of Arts in 1751, was ordained deacon at York in 1750, and became a

(1) *See* above, p. 149.
(2) Manuscript in the possession of the American Philosophical Society, Philadelphia, U.S.A.
(3) *See* Grigson (G.), *The Englishman's flora,* 147 (1955).

priest in Holy Orders in 1752. In 1753 he was appointed Rector of Keighley.

Knowlton referred in this letter to Collinson's orchids:

> 'I congratulate you much on y^{r} collection of Orchis wch are a prity tribe & to have y^{n} so elegant in a garden is somewhat extradinary, ye Goat [*Himantoglossum hircinum* (L.) Sprengel] is magnificent to a great perfection, it is many years since I have seen it, we have few in this country of note tho' have found ye green wing [*Orchis morio* L., Green-winged Orchis] in about 300miles from hence, but y^{r} son's zeal is somewhat extradinary.'

Himantoglossum hircinum is commonly called the Lizard Orchid, because the long twisted labella which hang round the flowering spike look like lizards. Several early botanists called this species 'Tragorchis', from Τράγος a goat, and orchis. In the words of the herbalist John Gerard in *The herball,* 160 (1597), 'the whole plant hath a rancke and stinking smell or savor like the smell of a goat, whereof it tooke his name.' When volume **31** of Abraham Rees's *Cyclopaedia* was published this species was still 'often called goat orchid'.[(1)]

Knowlton also mentioned several other plants grown by Collinson:

> '. . . y^{r} mountain magnolia [*Magnolia acuminata* L., Cucumber Tree, Mountain Magnolia] is very prolifick & ye Calacanthus [*Calycanthus floridus* L., Carolina Allspice] is extradinary I have y^{m} both, but presume will be long before either of y^{m} do so well with me. Hailica [*Halesia carolina* L., Snowdrop Tree] & ye accasia likewise, this last is finely in flower now [and] is a lovely tree but y^{t} Rhododendron [possibly *Rhododendron maximum* L., the Great Mountain Laurel or Rosebay, native of North America] is Inimitably fine & a capitall plant. Rair to be mett with, & more so in such a profusion of flowers, y^{r} vis [viz., namely] a pucken [*Sanguinaria canadensis* L., commonly called Puccoon, native of North America] is a raire plant unknown to me as is y^{r} Little thyme leaved sabina [*Juniperus sabina* L., Savin, native of the mountains of Central and southern Europe.] These Little Beautys deserves y^{e} admiration as minute ones are so to me.'

Rhododendron maximum was introduced into our gardens by Peter Collinson in 1736 and on 25 June 1756 he made a memorandum that

(1) Vol. **31** (1819) [1815] under 'Satyrium'.

Lizard Orchid (Himantoglossum hircinum (L) *Sprengel*), *native of south-west and central Europe and the Mediterranean region, is rare in England. Gerard, from whose* Herball *this plate is taken, mistakenly thought that more than one sort of plant was involved.*

read:[1] 'The Great Mountain Laurel, or Rhododendron, flowered for the first time in my garden'. Regarding *Sanguinaria canadensis,* it may be noted that all parts of this species have copious orange-red sap, whence its name. The Indians were said to paint themselves with the sap and the inhabitants of America called it by the Indian name 'Puccoon'.[2] Knowlton's letter to Collinson is dated 13 July 1767. Three months earlier Collinson had written to John Bartram in America about his *Sanguinaria,* telling him 'Puccoon ... and Claytonia [probably *Claytonia virginica* L., Spring Beauty, native of Virginia] in flower, April 5th.'[3] Collinson grew several species of Juniper in his garden. Knowlton was probably referring to the plant entered in Collinson's Catalogue as '*Juniperus sabina.* Sabina folio cupressi,

(1) Dillwyn (L. W.), *Hortus Collinsonianus. An account of the plants cultivated by ... Peter Collinson,* etc., 45 (1843).
(2) Miller (P.), *The gardeners dictionary ... The eighth edition,* etc., (1768).
(3) Darlington (W.), *Memorials of John Bartram,* etc., 288 (1849) Philadelphia.

berry-bearing Savin, Bauhin, Pinax 487.' According to Bean:(1) 'Although this is clearly a synonym of *J. sabina* L., the plants cultivated under the name are usually dwarf'. The whole plant emits a strong odour when bruised.

Knowlton continued by informing Collinson:

'Jn° Knowlton my nephew Gard^r at Chiswick will take care to send me ye print of adrachne [i.e. *Arbutus andrachne* L., commonly called Andrachne](2) as I will ord^r him to call some time or if Left for him roled up direct^d to him at Burlington-house he'l gett it safe Left with ye porture... Mr Bolton [possibly Thomas Bolton or his younger brother James](3) lately sent a Box of Cloudburrys [*Rubus chamaemorus* L., Cloudberry] but very poor tho one [on] Clump of ye earth... Last sumer I had some plants drawn up... from woodland more [moor] near chapell of frieth Darbyshire [i.e. Chapel-en-le-Frith](4) strong & fine so y^t Il try to gett some from thence if I can, this sumer. M^r [William] Constable[s] adrachne is in a pot so is mine & dare not trust y^m to our winters & cold slugish soyls... I have not heard of late from Mrs Blackburn nor ye esq^r [John Blackburn], ye portugal cypresse I now Keep in pots, as they will not abid[e] our winters, hear at no rate, as I have to my loss'

This tree named by Miller *Cupressus lusitanica*, and commonly known as Cedar of Goa, was introduced into England from Portugal during the latter part of the seventeenth century. It was, however, never found wild either in Portugal or the Portuguese settlement of Goa in western India, and it is now certain that it is a native of Mexico.(5) Knowlton was correct in saying that it could not abide Yorkshire, as it is not suited for planting in the north of England.

He continued:

'I do assure you Last winter was a saver [severe] one to me, yett have 41 sorts of ficoides [*Mesembryanthemum*], yett some few I much want and would give a good prise to have y^m if I did know where that I could gett y^m Y^r situation is extraordinary for perservation & more so to have such plentyfull cropes of fruit [Here] less was hardly ever known than this season: we have

(1) Bean (W. J.), *Trees and shrubs hardy in the British Isles,* **2**, 494 (1973).
(2) *See* above, p. 155 and footnote.
(3) *See* above, p. 245.
(4) Among places cited where *Rubus chamaemorus* is found growing in Derbyshire are Chapel-en-le-Frith and also about five miles from that locality at Jaggers Clough, near Woodlands Valley. *See* Clapham (A. R.), *Flora of Derbyshire,* 177 (1969).
(5) Bean (W. J.), *Trees and shrubs hardy in the British Isles... Eighth edition,* **1**, 803–4 (1970).

had most heavy rains for a long time so y^t^ we have wished for sun & fine weather...I am realy sorrey for M^r^ [John] Bartram['s] short allowance, however hope he'l Live to a good old age.'

The American botanist John Bartram, born 23 March 1699, near the village of Darby in Pennsylvania, developed an early interest in natural history. About 1730, he built himself a house by the Schuylkill river near Philadelphia where he established a botanic garden. He spent some forty years of his life collecting plants in North America, his journeys taking him to Florida, Georgia, North and South Carolina, New York State, Pennsylvania, and Virginia. During these years he sent shipments of plants and seeds to Peter Collinson, Lord Petre, Philip Miller, and other correspondents, which resulted in our gardens being greatly enriched with indigenous plants of eastern North America.

Through the help of Peter Collinson, Bartram, in 1765, was appointed botanist to George III with a salary of £50 a year.(1) In 1767 the botanist complained that this sum did not cover the cost of 'the freight and other charges' incurred in sending plants to the King.(2) Moreover, in June 1766 Bartram had written to Collinson to say that the collecting of specimens had cost him 'many scores pounds, pains, and sickness.'(3) Such financial difficulties may explain the meaning of the above reference to 'Mr Bartram['s] short allowance'. He died 22 September 1777 at the age of 78.

After complaining further of the severity of the winter by telling Collinson: '...all ye figs in this part of ye world sufferd much by ye winter so very little fruit on y^m^.' Knowlton left his letter unfinished until 14 July, when he wrote:

'...yesterday was sent a very smal Box to ye Hull carier to be deliverd to ye London York carier wch sends out one [on] fryday will be in London Saterday 7 night directed to you in Grace church street London: in it is a clump of ye dwarfe Honeysuckle [*Cornus suecica* L., commonly known as Dwarf Cornel], 1 plant of ye anagalis moonela [*Anagallis monelli* L.] & one of ye comon Blew as youl see & hope will come safe to y^r^ hands [this was possibly the variant of *Anagallis arvensis* L., Scarlet Pimpernel, with blue flowers]...have you seen at Mr Lees in flower ye Alstrameria wch is so

(1) Darlington (W.), *Memorials of John Bartram*, etc., 268 (1849) Philadelphia.
(2) *ibid.* 287.
(3) *ibid.* 281.

elegant & fine a flower, both m[r] Aiton [Gardener at the Royal Botanic Garden, at Kew] & he tells me so.'

The species which received such praise is *Alstroemeria pelegrina* L., commonly called Spotted-flowered Alstroemeria. This native of Peru was introduced into this country in 1753 by Kennedy and Lee of the Vineyard nursery, at Hammersmith, near London. About 1745 the two partners were Lewis Kennedy[(1)] and James Lee (1715–95). The firm of Kennedy and Lee became noted for their successful culture of exotic plants, and also for the many species introduced or first cultivated by the firm.[(2)]

Knowlton would certainly have liked to come up to London in order to see the Alstroemeria in flower. He seems always to have enjoyed the rare occasions when he came south. At these times he was able to meet eminent naturalists, botanists, and gardeners, and visit places of interest including gardens in, and around, the Metropolis. Of his liking for London he told Collinson:

'... you Live in y[e] centure of Knowledge & where I could wish to be but alas y[e] distance is such y[t] when I do come thire, hardly know how to part [from that] w[ch] has so many Inchanting pleasures to engaige me.'

Finally, Knowlton ended his letter by again writing of the hardness of the winter:

'I am very sorrey to hear y[t] thire is such a losse in y[r] plants at M[r] Thorpes & more so y[e] ficoides [*Mesembryanthemum*] I want but few to compleat mine... & am realy sorrey I cant be suplyd with those I want: y[e] winter was crowel [cruel] hear I doe asure you, yett I did perserve [preserve] y[m] notwithstanding y[e] frost did one night touch y[m] I have now I fear tired you out with this long epistle but conclude with esteem & Love y[r] very Humble servant

Tho Knowlton

P.S. y[e] flower fence continues flowering yearly at Chatsworth &c. July y[e] 17.'

(1) Considerable confusion appears to exist as to the identity of this Lewis Kennedy. *See* Harvey (J.), *Early nurserymen,* xiii, 84–5 (1974).

(2) Willson (E. J.), *James Lee and The Vineyard nursery Hammersmith,* (1961) Appendix.

TO JOHN ELLIS

Chatsworth, October 1770[1]
'Oh my Good S^{r}

I had ye Great pleasure to hear of y^{r} health by M^{r} [John] Blackburne & Misse [Anne Blackburn] where I had been for a few days [i.e. at Orford Hall, near Warrington, Lancashire] saw ye tea tree [*Camellia sinensis* (L.) O. Kuntze] for ye first time in Great good health with n^{os} of others almost as raire, but not y^{r} fly trap [*Dionaea muscipula* Ellis, Venus's Fly Trap, native of Carolina] tho he had no less y^{n} 5 of y^{m} I had on [one] so had Ld Exeter [Cecil Brownlow, eighth Earl of Exeter (1701–54), the owner of Burghley House, near Stamford, Lincolnshire, and a residence at Chelsea, near London] & Mr Brownrig of chelsey[2] [possibly William Brownrigg] but all suddenly Lost y^{m} so have seen it not to perfection; y^{r} gardainia [*Gardenia jasminoides,* Cape Jasmine] has these 2 years been naild up against a south wall & in such vigior as... not at first to Know it, haveing so shining &. florid apearance, I have it & ye single in y^{r} Little Book[3] &. have an other Likewise wch is a full flower & a smaller Leave diferant from eather wch yett I never see any where else, we had of it ye Butcher at Barnett [near London, named] Clark. I have seen ye uva ursi [*Arctostaphylos uva-ursi* (L.) Sprengel, Bearberry] found on ye mountains in Scotland & ye bettula nana [*Betula nana* L., Dwarf Birch] from y^{t} country wch I had not seen... before alive tho by ye faver of m^{r} White a surgeon in York [I] had a good specimen of it w^{ch} [he] asured me [he] geatherd thire. I have ye chamaepericlymenum or Dwarf Honeysuckle [*Cornus suecica* L.] found in ye Greatest plenty near Salters Gate [about seven miles north east of Pickering] & Thornton... 3 or 4 miles east [of Pickering, both in Yorkshire, in North Riding] wch I esteem as a very Great curiosity & [it] dos well with me as do many of ye raire English Plants. I shall in a few days go to ye Duk of portlands & norfolks

(1) Linnaeus corres. vol. 17 (Supplement) preserved in the Library of the Linnean Society of London. This letter published in 1821 in Smith (Sir J. E.), *A selection of the correspondence of Linnaeus,* etc., **2**, 78–80 (1821).

(2) William Brownrigg (1711–1800) practised medicine at Whitehaven, Cumberland. There is no record that he ever lived in Chelsea and, as Knowlton added 'Chelsey' in his letter as an after-thought, above 'Brownrig', it seems possible that the word was misplaced. According to Faulkner (T.), *An historical and topographical description of Chelsea,* etc., **1**, 64 (1829) Hannah Sophia, Dowager Countess of Exeter, the wife of the eighth Earl, was the owner, until she died in 1765, of a villa in Chelsea, called 'Chelsea Farm'. It seems likely, therefore, that Knowlton intended to write 'Chelsey' in his letter above 'Exeter'.

(3) Ellis's 'Little Book' may have been a reference to a *hortus siccus* or, perhaps, to a bound volume consisting of reprints of two accounts of the *Gardenia*: (1) An account of the Gardenia ('Cape Jessamine') by John Ellis, *Phil. Trans.,* **51**, Pt. 2. 932–35, pl. 23 (1761 for 1760), and (2) An account of the Gardenia by Daniel C. Solander. *Phil. Trans.,* **52**, Pt. 1, 654–61, pl. 20 ('of a Gardenia with a single flower') (1762 for 1761).

L^{d} scarbros & so home hope to see ye tea tree at y^{e} on [one] of y^{m} as you sent so many seeds to worksope. I could wish y^{t} our young Duk would Like his father take a turn y^{t} way whom every day Improved in Knowledge & y^{t} you was acquainted with him whom is so good & I hope amiable in all things.'

Each of the three peers, namely William Bentinck, second Duke of Portland (1709–62), Edward Howard, fourteenth Duke of Norfolk (1686–1777), and Richard Lumley-Saunderson, fourth Earl of Scarborough (d. 1782), were the owners of several estates, but it is likely that Knowlton's journey took him to Bulstrode, the Duke of Portland's seat near Beaconsfield in Buckinghamshire, the Duke of Norfolk's estate at Worksop, 26 miles north of Nottingham, and Lumley Castle in County Durham belonging to the Earl of Scarborough. In writing of the 'young Duke' Knowlton was referring to William Cavendish, fifth Duke of Devonshire (1748–1811) whose father, it will be remembered, had married the Earl of Burlington's youngest daughter. Knowlton continued:

'...it would be most singular pleasure to have a Lett^{r} of Information now & y^{n} from you as my worthy Lat[e] fraind m^{r} Collison is gone & no more to be seen [Peter Collinson died 11 August 1768]: bege it as a faver in you to Indulge an old man of 80 years whom is as Great a Lover perhaps as any now y^{t} is Liveing otherways would not take such a journey y^{t} I have done, & if y^{t} I could be of any use to you by a sketch of any stoves y^{t} I have seen any where I should be glad to oblige or please you in any...thing y^{t} is in my power for I do much esteem you for y^{r} writings & fraindship...I am now told y^{t} you are again about to publish wch please to Inform me, & what is ye subject as I could wish to Know & youl much please & oblige y^{r} fraind & very Humble sarvant
at Londesburght near York'
Tho Knowlton

As Knowlton asked Ellis to write to him at Londesborough, it appears that he was only on a visit to Chatsworth, from where he addressed his letter. In a postscript he wrote:

'PS I hope y^{r} nephew [Henry Ellis, F.R.S.] y^{e}...Governor [of Georgia] is well...I shall be pleased to hear that he is so, as his Hudson Bay History is much admired by me [i.e. *A voyage to Hudson's Bay...in 1746 and 1747, for discovering a North West passage: with... a short natural history of the country,* published in 1748]...

Earlier a quotation was given from a letter addressed by Thomas Knowlton to John Grimston of Kilnwick.(1) Further correspondence addressed to Grimston by Knowlton is given below.

TO JOHN GRIMSTON

20 Xber [December] 1775(2)
'Hon^d S^ir
...I hope y^t you have all Mr Penants works of Scotland, as they are very pleasing as well, as just, I am with much esteem & respect y^r most obedant Humble
Tho Knowlton'

Thomas Pennant published in 1771 his *Tour in Scotland,* describing the journey made by him in 1769. A supplement to this work was issued in 1772. In May 1772, Pennant undertook a second Scottish journey and this he described in the two parts issued in 1774–76 of *A tour in Scotland and voyage to the Hebrides.* On that occasion Pennant invited John Lightfoot (1735–88) to accompany him. As a result of the tour Lightfoot compiled a *Flora Scotica* which was published in 1777 at Pennant's expense.

TO JOHN GRIMSTON

11 Jan: 1777(3)
'Hon.^d S^ir
I hope y^t you have Recd my Lett^r sent... sometime since with a Lett^r of D. Costa [i.e. Emanuel Mendes da Costa]... I sent you ye cat: of Mr Perfetts with his Prises of oaks and other shrubs [possibly the *Catalogue of forest-trees, fruit-trees, evergreen and flowering-shrubs,* issued early in 1777, by the firm of nurserymen William and John Perfect in Pontefract, Yorkshire]... ye Bearer is a very sober man & atentive with ye utermost diligence y^t I am very sertain, & if it was in my power to doe him any service no man in ye world would be more willing, & if in y^r power bege it as a favor to Look one

(1) *See* above, p. 261.
(2) Grimston corres. preserved in the Humberside Record Office, DDGR 42/25.
(3) Grimston corres. DDGR 42/26.

[on] him as one deserveing of Incorragement; as hes Born in y^r^ naborhood... y^r^ most obedant
Tho Knowlton'

It is evident from a number of Knowlton's letters that he was always willing to help worthy and industrious young men to obtain work.

TO JOHN GRIMSTON

27 March 1777(1)
'Hon^d^ S^ir^
about 5 weeks agoe I wrot to you &...sent...a Lett^r^...from M^r^ Da Costa...begeing y^r^ Bon(2) of ye Yorkshire shells...you so Kindly oferd him...with many...hint & observation & hope y^t^ youll send y^m^ accordingly...I have mad a great plantation of shrubs & Mr Evans now take pains in his Garden so I was stretched to oblige him y^t^ now I am not able to part [with] any more [plants?]
I send you some mellon as you see by ye Inclosed I am with Esteem y^r^ obedant Humble
T. Knowlton'

It has been suggested to the compiler that 'Mr Evans' was the Revd. Robert Evans, Rector of Londesborough from 1769 to 1788.(3) It appears that Knowlton had made the plantation with shrubs from his own garden and that after he had done this his stock was so depleted that he had no more plants to spare.

Earlier reference was made to da Costa's *Natural history of fossils,* published in 1757. Soon afterwards the same author's *Historia naturalis Testaceorum Britannicae, or, the British conchology* appeared.(4) It was intended that copies of this work should be sold with illustrations consisting of water-colour drawings, uncoloured engravings, or engravings coloured by hand. According to the list of subscribers John Grimston wished for a copy with water-colour drawings. However, finally, the book could only

(1) Grimston corres. DDGR 42/27.
(2) Bon, a favour, a gift, a thing freely or graciously bestowed. *See* O.E.D.
(3) The compiler is grateful to Mr David Neave for this information, in *epist.* 27.ii.77.
(4) In the Department of Zoology in the British Museum (Natural History) is preserved a volume with 17 water-colour drawings on vellum; all except the third drawing are identical with the plates in *The British Conchology.* They are presumably the original drawings prepared for the published work.

be purchased with engraved plates, seventeen in all, these being available either uncoloured, or coloured by hand. This is evident from the following letter written by Thomas Knowlton to Grimston:

TO JOHN GRIMSTON

'July 24. 1779[1]

I Recd y^{r} former faver for wch I thank you, a few days ago I Recd a letter from M^{r} Da Costa informing me y^{t} his conchology is now finished & says his thoughts of a painted seat has not succeded... & desires to Know if Youl have a collourd att 1. 11. 6 as you had desired one of ye painted ones [at] 3 Geneys & half... y^{r} answer will be agreeable to S^{ir} y^{r} very Humble
Tho Knowlton
NB perhaps my son & grandson may wait one [on] y^{u} & with M^{r} Da Costas Lett^{r} of July ye 10th some Day next week if convenant'

From the wording of this letter it seems that *The British conchology* was not published until 1779, even though the imprint on the title-page is dated 1778.

Knowlton's letters to Grimston appear to mark the end of his correspondence. By that time he was a very old man who had not long to live; he died in November 1781. Furthermore, it is learnt from a notice in the *Gentleman's Magazine,* **9** (New series), 544 (May 1838), that towards the end of his life he became 'totally blind'.

(1) Grimston corres. DDGR 42/29.

Appendix A

Knowlton as director of the work done in the planning and planting of Sir Marmaduke Constable's estate at Everingham in Yorkshire

Sir Marmaduke Constable (d. 1746) of Everingham Hall in the East Riding, a member of the Roman Catholic church, was on the Continent from 1730 until 1743, and during these years his estate and business affairs were supervised by his Chaplain, Dom John Bede Potts (d. 1743). As in many other estates at this period, much tree-planting was carried out in the park and garden at Everingham – a project that was under the direction of Thomas Knowlton. A large number of letters written by Potts contain passages informing Sir Marmaduke about the progress of this work, while others by Constable give his views on what was being done as well as suggestions for further improvements. Below are given quotations from the letters on such matters as the work he was required to do, his opinions concerning various plans, the payment he received, and details of the trees that were planted. It is evident from these extracts that Knowlton was thought very highly of as a gardener and planter. Everingham was only about 3½ miles south west of Londesborough and he spent much time there when Burlington was not in residence.

Several of the original letters from Sir Marmaduke to Potts are still extant, as are the copies Potts made of all the letters he sent to, or received from, Sir Marmaduke. They are all preserved in the Library of the University of Hull. In the interest of consistency, the quotations below are all derived from the copies. In 1976 the Yorkshire Archaeological Society published volume **136** of their Record Series, entitled, *Constable of Everingham estate correspondence 1726–43*, a work edited by Professor Peter Roebuck. Some of the quotations given below first appeared in letters published by Professor Roebuck, but many others are from material that he had to omit for reasons of shortage of space.

Potts dated his letters according to the usage current in England until 1752, that is by the Julian Calendar ('Old Style') under which the year began on 25

March. Other countries in Europe, however, had adopted the Gregorian calendar ('New Style') with its year beginning on 1 January. Furthermore, under the Julian calendar England was eleven days behind the rest of Europe, but this was corrected in 1752 by adding eleven days and jumping from Wednesday 2 September to Thursday 14 September. In Great Britain and Ireland the change to the Gregorian calendar, New Style, was effected by 'Chesterfield's Act' (24 Geo. II, *c.* 23), passed in March 1751.

To assist readers, the letters given below are placed in chronological order, treating the year of all letters as having begun on 1 January. Any letters written by Potts, however, before 25 March, have both forms of dating given. Thus, a letter dated by Potts 7 February 1734 (O.S.) is placed in chronological order according to the year 1735 (N.S.), and is written thus, 7 February 1934/35. Allowance has been made for the eleven days lost. For example Sir Marmaduke's letter dated Dunkerque 21 February 1742 precedes by ten days that of Potts written from Everingham 20 February 1741/42.

TO SIR MARMADUKE CONSTABLE FROM DOM JOHN BEDE POTTS

Everingham, 2 November 1730[1]

'... About a fourtennight agoe Mr Lawson,[2] and Mr Knowlton came here to consult aboute ye leveling ye Hall Garth;[3] and this day Mr Knowlton came again, to strik ye levil; so that work is begun. He has sent us one who understands leveling, and who was formarly one of his Apprentices...'

TO SIR MARMADUKE CONSTABLE FROM DOM JOHN BEDE POTTS

Everingham, 30 January 1730/31[4]

'... As to y^r Questions I answer as justly as I can. Y^e first is concerning ye hall Garth. Before that work was begun I desired Mr Lawson and Mr Knowlton to come here and agree upon ye most proper, and cheapest way to doe, ye latter is ye only director, and to his judgement I referd ye whole work, since ye other never came but once after. I am sensible ye expence will be more than you would allow, but his designe is to doe it so, as to need noe amendment...'

(1) DDEV/60/86/p. 3.

(2) Nothing appears to be known of Mr Lawson. Was he perhaps employed at Moreby Hall, south of York, in East Riding? *See* p. 286 footnote.

(3) Garth: a small piece of enclosed ground, usually beside a house; a field, paddock, garden, yard; a guarded or fenced place.

(4) DDEV/60/86/p. 5.

TO DOM JOHN BEDE POTTS FROM SIR MARMADUKE CONSTABLE

Blois, 9 May 1731[1]

'...I would have a Line of Ellmes brought down from ye style goeing into ye Churchyard all along Jacksons Garth to ye back way leading up to ye house and sett with Quickwood well fenced from horse and beast and ye town Gate. Gett Mr Knowlton to sett y^m out for you as he shall best fancy... The boarders for ye fruit trees of each wall must be first dugg manured well, and y^t well rotten before ye trees are planted... My L^{ds} Gardner[2] will likewise assist you in this, and chouse you for y^r Ellm plants such as are most proper...'

TO SIR MARMADUKE CONSTABLE FROM DOM JOHN BEDE POTTS

Everingham, 14 May 1731[3]

'...The expences have been great beyond expectation but ye work being carried on by Mr Knowltons directions and often personall presence who is a Master of that art I hope it will please...'

TO DOM JOHN BEDE POTTS FROM SIR MARMADUKE CONSTABLE

Blois 28 May 1731[4]

'...As I directed a row of Ellms to be planted down Jacksons Garths from ye Church stile to ye back way to my house, I believe you may find some other Garths y^t are bare to ye street, y^t will be proper to plant too... If you plant any in ye Park, I would have you put togeather att least 200, or 150 in each parcel. Mr Knowlton will find you proper places and soil. Keep from before ye house, East West, and South; any where else where you and he shall think proper, I shall approve off...'

TO SIR MARMADUKE CONSTABLE FROM DOM JOHN BEDE POTTS

Everingham, 29 May 1731[5]

'...I hold y^n before y^t ye boarder in ye Hall Garth was finished, but is not as you direct menured, I proposed that to Mr Knowlton, but as much as I remember he is against menuring of it since it is fresh ground...'

(1) DDEV/56/30/p. 120. Blois: a city in France, on the Loire, 100 miles south west of Paris, and capital of the department of Loire-et-Cher.
(2) i.e. Thomas Knowlton.
(3) DDEV/60/86/p. 9.
(4) DDEV/56/30/p. 122.
(5) DDEV/60/86/p. 11.

TO SIR MARMADUKE CONSTABLE FROM DOM JOHN BEDE POTTS

Everingham, 18 June 1731[1]
'... As soon as I see Mr Knowlton I shall acquaint him with y^r designe of planting ye Garths, and take his direction...'

TO DOM JOHN BEDE POTTS FROM SIR MARMADUKE CONSTABLE

Paris, 23 July 1731[2]
'... As to ye planting out of ye Ellms from Scorbrough[3] I leave to you and Mr Knowlton... You must measure over ye two South walls of ye Court before ye house and as many four yards as they measure so many fruit trees I will send you. The Pears you must plant where ye wall is ye highest, and ye ground ye strongest... I buy my trees from ye Carthusians[4] here who are supposed to have ye best collection of France...'

TO SIR MARMADUKE CONSTABLE FROM DOM JOHN BEDE POTTS

Everingham, 2 August 1731[5]
'... Mr Knowlton was here thursday last to inoculate some stock w^{ith} y^r old Apricock and Peach buds, and has inoculated to ye number of 19 stocks. I told you before y^t he had planted out two stocks wch he inoculated last year in ye east boarder in ye hall Garth, and they toke and flourish to admiration. He thinks it impossible to remove so great a quantity of trees as you have at Scorbrough in one season, and therefore advises to divide y^m into three, and to draw the strongest first. He will meet Mr Lawson here, who is he says, and I believe knows more of y^r designe in that kind, than any else. However we have resolved to plant in ye park in parcels as you direct...

TO SIR MARMADUKE CONSTABLE FROM DOM JOHN BEDE POTTS

Everingham, 10 October 1731[6]
'... Ther is no preparation as yet made for planting ye Ellms to be brought from Scorbrough because Mr Knowlton would have Mr Lawson present, as knowing

(1) DDEV/60/86/p. 12.
(2) DDEV/56/30/pp. 123–24.
(3) Scorbrough or Scorborough: a parish and small village in East Riding, about four and a half miles north north west from Beverley.
(4) Carthusians: a branch of the Benedictines, a religious order instituted by St Bruno in the eleventh century. The word is formed from Carthusianus, or Carthusiensis, a denomination given them in Latin, from a village in Dauphiné, called Chartreuse, where the first monastery of this kind was erected.
(5) DDEV/60/86/p. 15.
(6) DDEV/60/86/p. 18.

more of y^{r} minde in that kinde than any besides, they made two appointments to meet here, but att each time ye one was gone before ye other came, so nothing was effected. I shall endeavour to have y^{m} here as soon as possible. But as to y^{r} sending fruit trees from that side of ye sea it is ye universal opinion y^{t} they will not doe here...'

TO SIR MARMADUKE CONSTABLE FROM DOM JOHN BEDE POTTS

Everingham, 4 January 1731/32[1]

'I writt to y^{u} before y^{t} it was resolved upon to plant part of ye Ellems in ye Park: but three weeks agoe Mr Lawson and Mr Knowlton meeting here to consult upon that affair they concluded to ye contrary nor to plant out any in parcels, or Clumps till y^{r} return, and only to plant in ye Garths and ye street a line on west side begining at Allerthorps Garth and so on in a line, and from ye church to ye highway: Nor will Mr Knowlton sett out in ye Hall Garth two boarders for hedges, unless you order it a new, alledgeing y^{t} it will be much better to stoop[2] and rail[3] ye walls, than to have a hedge: but if y^{u} persist in y^{r} resolution he will then sett out ye boarder, and pitch upon such plants as he shall think fitt. He has commanded the two plots on each side of ye walk to be dug 2 spit deep to plant ye trees y^{t} are to be brought from Scorbrough. I wonder you would buy such a quantity att once... Mr Knowlton will sett ye trees himselfe when they come; and if you have a minde for any beach plants [*Fagus sylvatica* L., Beech], he will give you some, haveing more than he will have need for. The weather is now open, and if it continue so shall send for two loads of Ellems ye next week ye ground being prepared...'

TO SIR MARMADUKE CONSTABLE FROM DOM JOHN BEDE POTTS

Everingham ?February 1731/32[4]

'...On ye 16 of January I had notice of ye fruit trees being come to Hul, and sent for y^{m} ye next day. Mr Knowlton came to take y^{m} out of ye box found y^{m} to have been carefully put up, but some what dry, and therefore steeped y^{m} in water new milk warme, and than put y^{m} into a cellar packe up in very wett staddle hay[5] to moistem y^{m} more, and preserve y^{m} from frost ye weather than being very uncertain, he likes ye Peaches, but not ye Pears, nor some others, being of five or six years standing; He set every tree himselfe in ye two new boarders prepaired for y^{m}, but there being not room was obliged to plant some of ye Pears against ye side of y^{e}

(1) DDEV/60/86/p. 21.
(2) Stoop: to mark out with stoops or posts.
(3) Rail: a horizontal bar of wood or metal, fixed upon upright posts as part of a fence.
(4) DDEV/60/86/p. 23.
(5) Staddle hay: a portion of a stack begun and left unfinished on account of wet or other causes, *See* Wright (J.), *The English dialect dictionary*, etc., **5**, 715 [vol. 1–6, 1898–1905].

house in ye hall Garth... All ye Ellms at Scorbrough are brought home payd for and planted as I told y^{u} in my last. Mr Knowlton sett every tree, was three whole days here and very often last winter and this too ordering ye leveling and is ready upon ye first notice to assist in every kind. I presumed to give him two guinas: I know not how you'll take it, but it is short of what he deserves. He has sent you 1500 beach seedlings [*Fagus sylvatica* L.] wch are sett in ye Garden. If you will have two hedges in ye hall Garth, he says ye beach will be more proper than Ellm: but he is against both: and have ye boarder stooped wth a single rail...'

TO DOM JOHN BEDE POTTS FROM SIR MARMADUKE CONSTABLE

Cette on ye Meditaraneum Sea,
26 April 1732[1]
'... Reward Mr Knowlton hansomely according to his merite, and you are ye best judge of y^{t}, who are present. I agree with him about ye Hall Garth, lett it be stooped raild and painted...'

TO SIR MARMADUKE CONSTABLE FROM DOM JOHN BEDE POTTS

Everingham, 24 June 1732[2]
'... Mr Knowlton was here yesterday, and has nailed all ye young trees upon ye two new boarders, two of ye 80 are dead, and one as yet in a dubious way ye trees dead are plumbs...'

TO SIR MARMADUKE CONSTABLE FROM DOM JOHN BEDE POTTS

Everingham, 9 January 1732/33[3]
'... Mr Lawson, and Mr Knowlton meet here about ye middle of ye last month: they then thought it convenient to plant out all ye beds into Blackburn garth, that they may have more air, strick larger roots, and be in less danger when they are planted out into ye Intack[4]...'

(1) DDEV/56/30/p. 125. Cette: a town in Languedoc, France, 17 miles south west of Montpellier.
(2) DDEV/60/86/p. 27.
(3) DDEV/60/86/p. 35.
(4) Intack, intak or intake: a piece of land enclosed from a moor, common, or road: land reclaimed from a river.

TO SIR MARMADUKE CONSTABLE FROM DOM JOHN BEDE POTTS

Everingham, 2 March 1732/33[1]
'...The underwood in the Plantation is renewed a fresh. Mr Knowltons men are cutting ye wall trees. he was here yesterday and cut all ye young trees y^t came from Paris himselfe, no more dead than what I acquainted you off before...if no blast or storme come I hope to tast of ye fruit. Mr Knowlton has sent you two thousand fur seedlings of two years. youll be att more expences, than you have been higher too in Keeping plants clean from weeds both in ye Garden and Blackburn Garth wch is now in greet measure planted...The Intack is to be...ploughed twice or thrice...ye it may be fitt to remove ye strongest plants out of Blackburn Garth. Mr Knowlton and I do not take your directions alike as to ye fosse[2]. He thinks y^t ye fosse must be no longer than ye front of ye house butt I that it is to be ye length of ye Garden wall...I transcribed that part of y^r letter and gave it him, who will consider of it better and be here in 14 days time again to sett out ye fosse, I know he will do it to ye best advantage...'

TO SIR MARMADUKE CONSTABLE FROM DOM JOHN BEDE POTTS

Everingham 12 April 1733[3]
'... Mr Knowlton has sent 1000, more of furr seedlings, planted in Blackburn Garth; but harrisons hens cratches partt up...'

TO SIR MARMADUKE CONSTABLE FROM DOM JOHN BEDE POTTS

Everingham, 8 May 1733[4]
'...I gave you account before y^t your strongest Ellms in ye garden, and most of ye oak and Ash [probably *Fraxinus excelsior* L.]...raised there were removed into Blackburn garth by Mr Lawsons and Mr Knowltons directions, there reasons I gave you in a formar letter. The next week Robt[5] says he will order ye ploughing of ye Intack. I desired twice to know how you would have it fenced of [off] from its neighbours, ye same I here desire... Mr Lawson and Mr Knowlton were here yesterday and have staked out ye ditch. The wall of ye Garden is to be lengthened 26 yards...wch brings ye ditch to an equal square with ye house. I must acquiesse; but do not take this to be your meaning when you ordered ye East and West wall, to be carried as far out as ye ditch – they say it must be done as staked out and that you do not reflect y^t ye East corner wall goes so far out...'

(1) DDEV/60/86/pp. 36–37.
(2) Fosse or foss: a trench, a bank with a ditch.
(3) DDEV/60/86/p. 39.
(4) DDEV/60/86/pp. 39–40.
(5) i.e. Robert Usher, who served as steward of the Yorkshire estate of Everingham, under the general supervision of John Bede Potts.

TO SIR MARMADUKE CONSTABLE FROM DOM JOHN BEDE POTTS

Everingham, 8 June 1733[1]
'... Mr Lawson and Mr Knowlton y^{t} are readie at ye ditch, [it] must be done as described in my last, dated ye 8 of May, and if don otherwisse will be very irregular...'

TO SIR MARMADUKE CONSTABLE FROM DOM JOHN BEDE POTTS

Everingham, 7 July 1733[2]
'... Mr Lawson is with me and sends his service to you and is positive y^{t} ye ditch must be drawen as sett out by Mr Knowlton or else ye wall will be more crooked than before...'

TO DOM JOHN BEDE POTTS FROM SIR MARMADUKE CONSTABLE

Luca, 3 November 1733[3]
'... Mr Lawson and Mr Knowlton takes my designe of altering ye Garden wall perfectly right, and am sure ye Latter can do it, better y^{n} I... therefore lett him proceed in y^{t}, as he shall judge best. Proceed allso to plant ye Intacke as Mr Knowlton shall think convenient... Cutt up all ye decaying plumb trees, plant others of ye best you can gett, or cherries. Be often makeing Mr Knowlton a handsome present answerable to what he deserves...'

TO SIR MARMADUKE CONSTABLE FROM DOM JOHN BEDE POTTS

Everingham, 5 December 1733[4]
'yrs from Luca of ye 3^{d} of Nov: I received ye 27... as to ye work begun in ye Intack... before ye receipt of y^{rs}, I had left ye whole management of that place to Mr Knowlton, nor indeed would he have had any thing to do with it unless he had had his own will... I had given to Mr Knowlton after he had planted ye wall trees, those y^{t} came from Scorbrough, and directed ye leveling of ye hall Garth and ye two boarders five guineas and since ye receipt of y^{r} last 3. gui. He is frequently here, and upon ye least notice is ready to come...'

(1) DDEV/60/86/p. 41.
(2) DDEV/60/86/p. 43.
(3) DDEV/56/30/p. 137. Luca i.e. Lucca: a city in Tuscany, Italy, ten miles north east of Pisa.
(4) DDEV/60/86/pp. 53–54.

TO SIR MARMADUKE CONSTABLE FROM DOM JOHN BEDE POTTS

Everingham, 21 January 1733/34[1]

'In my last I told you y^{t} Mr Lawson would make you a present of some large oaks, and about 400 Limes if you would be att ye charge wch is done, and were gott home just before Christmass and layd in ye ground, least a frost should prejudice y^{m} dureing ye holy days... Mr Lawson and Mr Knowlton were here ye 7th of this, and stayd to 11 at night. they both urge to have 300 scot[c]h furr [*Pinus sylvestris* L.] [£]33.15.0, 1000 beach [*Fagus sylvatica* L.] and 100 spruce [probably *Picea abies* (L.) Karsten] of each bought. for they propose to leave some quantity of Elms, to place where you please, and not to be sett in spungy or wett soil. I was much against it and talked whether they thought to plant all ye wood in yorkshire in that little compass, but if it be a [£] 5 or [£] 6 matter I shall not contradict y^{m}...'

TO SIR MARMADUKE CONSTABLE FROM DOM JOHN BEDE POTTS

Everingham, 27 February 1733/34[2]

'... ye planting in ye Intack is carried on att present more vigerously, because ye person y^{t} commands ye labourers under Mr Knowltons directions is to go up to Mr Duncome[3], so I hope three weeks as before mentioned in my former Letter will be finished. Mr Knowlton has made you a present of 3000 plants of different sort to ye value of £10 but as you were att ye charges of takeing up Mr Lawsons trees, so are you here... Mr Knowltons men are now cutting ye wall trees wch are in bloom and there are severall blooms upon those you sent over...'

TO SIR MARMADUKE CONSTABLE FROM DOM JOHN BEDE POTTS

Everingham, 29 June 1734[4]

'... we had a good many cheriyes but not one could be preserved till ripe being devoured by ye magpys, Jackdaws, and smaller birds wth wch we abound, tho I am apt to believe, y^{t} birds without wings has been more faulty... Mr Knowltons servant is now cutting all the trees, there is some peaches and nectarines upon ye young trees you sent over, but no fruit in any of ye others...'

(1) DDEV/60/86/p. 54.
(2) DDEV/60/86/pp. 57–58.
(3) Thomas Duncombe (d. 1746) of Duncombe Park, an estate within a mile of Helmsley in North Riding, Yorkshire. In 1736, it will be remembered, Knowlton wrote to Samuel Brewer telling him, 'I imagine I shall be sent for to Mr Duncombs... very soon they I know are Drawing to a conclusion of what I had... seat out.'
(4) DDEV/60/86/p. 47.

TO SIR MARMADUKE CONSTABLE FROM DOM JOHN BEDE POTTS

Everingham, 7 February 1734/35[(1)]

'... As for planting this spring I know not what can be don, till I speak to Mr Knowlton who I daly expect: we must first supply ye place of those that are perished in ye Intack ye far greater part of ye oaks Mr Lawson gave you are dead...'

TO SIR MARMADUKE CONSTABLE FROM DOM JOHN BEDE POTTS

Everingham, 8 April 1735[(2)]

'... Mr Knowlton hired two gardners from Beverly to cut ye trees, wch being now numerous takes up a considerable time, y^t ye Guinea you halfe yearly allowed him did not pay there wages, so told Robt[(3)] to pay y^m, and above y^t gave Mr Knowlton 2 guineas to make good what he had before layd out on this account, as allso for his comeing severall times here when desired. Those gardners thought fitt to cut down moste of ye old trees to witt pears, plumbs, and morel cherry to spring again, being mostly worn out: I wish it had been done ye year you went from home. The oaks y^t dyed last year in ye Intack are renewed, and more underwood sett in ye same place; but nothing planted a new, because no ground could be prepaird after ye receipt of y^{rs}, and ye rain so daly, y^t labourers could not stand it... Mr Knowlton by consent of my Ld Burlington has stocked y^r litle pond in ye Intack with fish viz. 18 Carp and 50 brace of pearch...'

TO SIR MARMADUKE CONSTABLE FROM DOM JOHN BEDE POTTS

Everingham, 14 May 1735[(4)]

'... I acquainted Mr Knowlton wth your resolution of planting in y^r piece of ground wch lyes paralel to ye Intack, but he says there is no thinking of planting there, unless it be pailed off, or ye Intack pails[(5)] removed farther into ye Park; wch I imagine would be a great disadvantage to what is planted already in ye Intack: besides he says y^t ye work in ye Intack must be continued a year longer... you may perhaps, imagine that ye whole Intack is planted but it is not, only as far as ye Birch spring[(6)]... There are four wall trees in ye Garden towards ye Intack y^t are sickly, one of wch Mr Knowlton has cut down within a foot or there abouts of ye grou-

(1) DDEV/60/86/p. 51.
(2) DDEV/60/86/pp. 55–56.
(3) Robert Usher, *See* above, p. 281 footnote.
(4) DDEV/60/86/p. 46.
(5) Pale: also written pail in East Yorkshire, a stake, especially as used with others to form a fence.
(6) Spring: a copse, grove; a young wood.

nds, a great many in all ye boarders has been blasted this spring... I think every Pear tree, and Plumb tree bears this year...'

TO SIR MARMADUKE CONSTABLE FROM DOM JOHN BEDE POTTS

Everingham, 23 August 1738[1]
'... The Elms must of necessity be taken out of Blackburn Garth, and transplanted where they are to stand: and Mr Knowlton who was against planting any more, till you saw what has been done, and how you Like that wch is planted in ye Intack, is alltogither for plenting y^{m} out, and says they will be spoilt if they stand any longer...'

TO MARMADUKE CONSTABLE FROM DOM JOHN BEDE POTTS

Everingham, 20 November 1738[2]
'... Robt[3] desired Mr Knowlton to come down and give his directions how to prepare y^{t} little plot of ground between ye birch spring and ye high road, but has not a minde to engage any more till he hear how you approve of y^{t} wch is done; yet I am apt to believe he would come down if ye family was not att Lonsbrough, wch has been there some time, and will continue as reported till after Christmas, he being much confined at home on y^{t} account...'

TO SIR MARMADUKE CONSTABLE FROM DOM JOHN BEDE POTTS

Everingham, 26 January 1738/39[4]
'... There is little planted for my Lord Burlington wch his family being still att Lonsbrough Mr Knowlton has not had ye liberty to come here to give his directions about that piece of ground between ye birch spring and ye high road, so y^{t} lyes undone, wch was designed to be prepaird and planted...'

TO DOM JOHN BEDE POTTS FROM SIR MARMADUKE CONSTABLE

Dijon, 27 February 1739[5]
'... I hope as you remove ye grown trees out of your Nurseries, you do not faile of supplying there places wth younger, so that your Nurserie ground allways remain

(1) DDEV/60/86/p. 84.
(2) DDEV/60/86/p. 86.
(3) i.e. Robert Usher *See* above, p. 281 footnote.
(4) DDEV/60/86/p. 87.
(5) DDEV/56/30/p. 148. Dijon: a city about 150 miles south east of Paris, and capital of department of Côte-d'Or.

full. I am sorry Mr Knowlton could not assiste you as usual. I hope you continue to reward him very handsomely for what he does for you. I could be glad to have had y^{t} plantation finished to ye high road, and do it well when you do it...'

TO ROBERT USHER FROM SIR MARMADUKE CONSTABLE

[Douay] 23 June 1739[(1)]
'... What ever Mr Knowlton proposes lett it be done. I suppose a good part of ye grass part of Ralphs garth will come into ye plantation, as allso ye sandy part of ye Pecker. Further I shall be obliged to alter ye division of ye two south Closes[(2)]; for I would carry an Avenue down to ye Beck[(3)] in ye Carr,[(4)] Give my service to Mr Knowlton and lett him know as much. I believe ye hedge of ye New-Close as far as ye Beck must allso be altered...'

TO SIR MARMADUKE CONSTABLE FROM DOM JOHN BEDE POTTS

Everingham, 23 April 1740[(5)]
'... Mr Lawson of Mowerby[(6)] who was here last week... gives his humble service to you, and has made you a present of 1400 furr seedlings two years old, wch are planted in a piece taken in on purpose, and hedged about to preserve y^{m} from ye hares, and Rabbits between ye Birch spring and ye high road; Mr Knowlton judgeing it more proper, to plant y^{m} some where else then in ye Garden, out of wch by his directions are transplanted Scotch furrs [*Pinus sylvestris* L.] of seven years grouth, and were his gift. He has been down three, if not fouer times... this winter...'

TO SIR MARMADUKE CONSTABLE FROM DOM JOHN BEDE POTTS

Everingham, 15 January 1740/41[(7)]
'... Mr Knowlton was here this day and since it is y^{r} minde y^{t} ye Birch spring should be planted, y^{t} is resolved upon, at least ye lower part of it... we have a good

(1) DDEV/56/30/p. 152. Sir Marmaduke did not state in this letter where he was staying at the time of writing, but as a letter dated 22 June 1739 is addressed 'Douay', it has been assumed that he was still in the same town on 23 June. Douay or Douai: on the river Scarpe, 18 miles south of Lille, France, and capital of an arrondissement in the department of Nord.
(2) Close: an enclosure, a place fenced in.
(3) Beck: a brook, small stream, or river.
(4) Carr: wet boggy ground, a fen or bog grown up with low bushes, willows, alders, etc.
(5) DDEV/60/86/p. 96.
(6) A reference possibly to Moreby in the East Riding of Yorkshire, five and three quarter miles south of York, where was situated Moreby Hall. Was this Mr Lawson the same as the man of that name who assisted with Knowlton in the planting of Everingham, and was he gardener at Moreby Hall?
(7) DDEV/60/86/p. 119.

quantity of Elmes, young oaks, plain trees, hassel [*Corylus avellana* L., Hazel]. But for ye low part of ye Birch spring oak must tak ye best...'

TO SIR MARMADUKE CONSTABLE FROM DOM JOHN BEDE POTTS

Everingham, 15 March 1740/41(1)
'...Since ye 20th of Jan. last that part of ye Birch spring from ye Pails, to ye hill where ye clump of birch stood, beyond ye dran [drain] is dug two foot deep, and planted, wth oak, hassels, secamoors [*Acer pseudoplatanus* L.], Elders, Birch, Poppler, Ossiers [*Salix viminalis* L.] and sarvers [Service Tree?]... All those of your own raiseing. From ye place called ye Pecker Leases(2) to ye place called oakorn Close is planted wth Scotch furs to ye number of 900, and more, they taking well in ye hilly part, and even grows beyond expectation, where other wood will not, and because we have not enough to fill up y^t place Mr Knowlton who has been here five times since he sett out y^t work, and when we begun to plant came soon in ye morning and staid to five oclock at night, directing and as busy as any of ye worke men, advises to buy 4 or 500 more furs to wch I condescended, and are bought, and will be brought home ye second of April, and will fill that place...'

TO SIR MARMADUKE CONSTABLE FROM DOM JOHN BEDE POTTS

Everingham, 14 June 1741(3)
'...There are some beach [*Fagus sylvatica* L.] wch were Mr Knowltons present in seedlings, we sett many of y^m and they that lived, came slowly on at first, but now they do very well for there are a great many of y^m about 12 or 14 foot high...'

TO SIR MARMADUKE CONSTABLE FROM DOM JOHN BEDE POTTS

Everingham, 20 October 1741(4)
'...As to ye commissions you gave me...a third commission was to gett Mr Knowlton to view ye road at Thorp but he being at his Lords seat in ye south(5), I must referr his information to his returne, as allso ye ploughing of ye Intack...'

(1) DDEV/60/86/pp. 120, 179. Page 179 has been misplaced, it should have been numbered p. 121 and followed p. 120.
(2) Lease: pasture, pasturage or meadow-land.
(3) DDEV/60/86/p. 182.
(4) DDEV/60/86/p. 121. Page 121 is mis-numbered and misplaced. It should have followed page 182 which contains the beginning of this letter.
(5) Possibly Burlington House or Chiswick House.

TO DOM JOHN BEDE POTTS FROM SIR MARMADUKE CONSTABLE

Cambray, 16 December 1741[1]

'... What is y^{r} stock of seedlings or young trees. have you tryed Beach, and Chestnut. The best tree is allso ye best underwood. In this country they plant Oak six, or seven foot high drawen out so from these woods, and they do extreemly well. How easily coud I fill my Park wth Oak and at what litle expence, if they woud thrive so wth us, in this way of planting. All about Antwerp ye country is sandy barren ground, yet I never saw Oakes grow better, and all planted in ye manner I here mention. No Nurserys in those parts for Oak all drawen from woods y^{t} are planted... Consult wth your friend Mr Knowlton in this affaire...'

TO DOM JOHN BEDE POTTS FROM SIR MARMADUKE CONSTABLE

Dunkerque, 21 February 1742[2]

'... I am glad you have gott ye Ash-trees cut down in ye pannel[3]. Propose to Mr Knowlton to make a green walk all along ye two ponds on each side. If so w^{n} he plants he must leave sufficient roome. I have no Idea left of Everingham so lett him do every thing as he judges best. W^{n} you have gott your walk finished to ye road y^{t} goes to Ha[r]swell;[4] I would then have Mr Knowlton tell me if he thinks it proper to carry ye same as far as ye beck in ye New Close, or Carr. The rest of what is to be done I leave to your discression and his...'

TO SIR MARMADUKE CONSTABLE FROM DOM JOHN BEDE POTTS

Everingham, 20 February 1741/42[5]

'... By ye first opportunity I shall acquaint Mr Knowlton wth your proposalls about a green walk in ye Pannel, and of carrying on ye Avenue to ye Beck in ye new Close Carr...'

(1) DDEV/56/30/pp. 163–64. Cambray or Cambrai: a town on the river Scheldt, 32 miles south south east of Lille, France, in the department of Nord.

(2) DDEV/56/30/p. 164. Dunkerque: the French name for Dunkirk, a seaport and town of France, department of Nord, in the Straits of Dover.

(3) Pannel or panel: the word was perhaps used here to mean that the ground resembled a panel in shape and relation to the surrounding area.

(4) Harswell: a parish in East Riding, Yorkshire, three and a half miles south west of Market Weighton.

(5) DDEV/60/86/p. 115. Page 115 is mis-numbered and misplaced. It should have followed page 124 which contains the beginning of this letter.

TO DOM JOHN BEDE POTTS FROM SIR MARMADUKE CONSTABLE

Dunkerque, 5 March 1742[1]
'...I should think ye Pannels woūd make a very good Nursery, and if it was dressed up from all rubish woūd contain a great many trees easily fenced upon ye many ponds y^{t} are in it, what will Mr Knowlton say to this lett me know. What ever is don lett him provide to plant as much of ye situation will allow to defend ye house against ye North winds to wch its now greatly exposed by cutting down ye old Ash trees...'

TO SIR MARMADUKE CONSTABLE FROM DOM JOHN BEDE POTTS

Everingham, 8 March 1741/42[2]
'...I communicated y^{r} designe of haveing a green walk all along ye two ponds to Mr Knowlton, but his answer was, that there is time enough to think of y^{t}, since ye finishing of ye walk to ye high road will take up this season...there being a great hollow by ye Beach spring he has procured for you 2000 Beach setts [*Fagus sylvatica* L.], and hornbeam [*Carpinus betulus* L.] to face ye walk about 3 years old, or perhaps better. I have not proposed y^{r} scheme, for making ye Pannel a nursery, since you'l see him before that can be medled wth...'

TO DOM JOHN BEDE POTTS FROM SIR MARMADUKE CONSTABLE

St Omers, 14 April 1742[3]
'...I had lately a Letter from a friend of mine at ye request of my Ld Burlington about continuing his avenue thro my Ground I shall most certainly comply with any thing my L^{d} requests y^{t} is in my power: I therefore desire you will upon ye receit of this send to Mr Knowlton, and lett him informe you w^{t} grounds, and in what manner my L^{d} woūd have ye said avenue to pass mine and where, ye same is to take york high road. Pray do not forgett this in y^{r} next to me...'

TO DOM JOHN BEDE POTTS FROM SIR MARMADUKE CONSTABLE

Paris, 21 June 1742[4]
'...I shall comply with what Ld Burlington has requested, and between you and me have spoke to Coll: Moysier[5] to adjust that affair with his Lordship...I am

(1) DDEV/56/30/p. 165.
(2) DDEV/60/86/p. 116.
(3) DDEV/56/30/p. 166. Saint Omer: a town on the river Ar, 22 miles south east of Calais, in the department Pas-de-Calais.
(4) DDEV/56/30/pp. 166–67.
(5) James Moyser Esq. of Beverley.

obliged to Mr Knowlton for ye trouble he gives himself in my affairs. I do not doubt but you make him a consideration from time to time as you think he deserves. You are ye only judge of that being present. I want to know if he approves ye planting of y^t part I have ordred, or woūd recommend some other. I own its all to ye south of ye house, and for that reason ye less wanted there as to shelter, but as its bad ground, and what yet has produced large oak timber, and why it should not do so in future time is to me beyond dispute. Little Jack Williams who now is here tells me ye wood in ye Intack comes on very well, as does every thing you have planted. Pray talk to Mr Knowlton and Mr Lawson, to ye latter my humble service if they can find out a piece of ground more or less about ye same size that they think more convenient. I should be glad to plant farther from ye house, but apprehend much planting woūd not be so easily kept in order...'

TO SIR MARMADUKE CONSTABLE FROM DOM JOHN BEDE POTTS

Everingham, 24 February 1742/43[(1)]
'... The south boarders in ye green, and kitchen Garden are renewed w^th... earth and the walls planted with young trees gott from york by Mr Knowltons directions...'

TO SIR MARMADUKE CONSTABLE FROM DOM JOHN BEDE POTTS

Everingham, 17 March 1742/43[(2)]
'... Mr Knowlton sent last night a quantity of Beech mast enough to plant ye Park if they take...'

Potts died 21 June 1743. He wrote only a few more letters and none contains any mention of Knowlton.

(1) DDEV/60/86/p. 190.
(2) DDEV/60/86/p. 199. The beginning of this letter is misplaced and numbered p. 190. It should have been numbered p. 198.

Appendix B

Knowlton the bibliophil

Knowlton was a book lover. This is evident from his correspondence and from the number of works published in instalments to which he subscribed. Although the books he read and studied were mainly botanical and horticultural, his interests covered various other branches of natural history as well as subjects as diverse as antiquities and local history. It may be assumed that he built up a considerable personal library which later formed the nucleus of the large and valuable collection of books owned by his grandson Thomas Knowlton the younger (1757–1837), who for many years was steward or agent in the service of the fifth and sixth Dukes of Devonshire. From a letter dated 10 August 1820[(1)], it is learnt that at one time Thomas the younger wished for a house where he could keep the books and property that had been bequeathed to him. He presumably inherited his grandfather's books and also those of his father, the Revd. Charles Knowlton, who died in 1814.

The library of Thomas Knowlton the younger was sold by auction on 12 February 1846 and the two following days, by L. A. Lewis at 125, Fleet Street, London. A study of the 748 entries in the *Catalogue* of this sale[(2)] reveals that the greatest number of books were on natural history subjects including many important herbals, works on botany, conchology, ornithology and zoology published in England and abroad from the sixteenth to the early nineteenth centuries. But there were also many other books such as Bibles and dictionaries, and books on architecture, antiquities, gardening, history, local history, theology, topography, and travel. Literature was not well represented. Many titles of works mentioned by Knowlton in his letters to Samuel Brewer and other friends are found in this *Catalogue* as well as those of the various works in which his name appears as a subscriber.

(1) *See* letter addressed by Thomas Knowlton the Younger to the Sixth Duke of Devonshire, dated 'Darley 10 August 1820', and now preserved at Chatsworth.

(2) A copy of this *Catalogue* is preserved in the Department of Botany in the British Museum (Natural History).

Among the many entries are some important botanical publications that appeared after the death in 1781 of Thomas Knowlton the elder. They include, Aiton's *Hortus kewensis* 5 vol. 1810–13, Bolton's *History of fungusses growing about Halifax* 3 vol and Appendix 2 vol. 1788–91, Curtis's *Botanical Magazine* 64 vol. (including General Index to vol. 1–53) 1787–1836 and Sowerby's *English botany* 36 vol. 1790–1814. Such works were probably acquired by Thomas Knowlton the younger who, like his grandfather, was devoted to the study of botany and natural history.(1)

Thomas Knowlton the elder must have attached special value to books given to him by eminent friends. Three of such gifts are publicised in the above *Catalogue,* as follows:

Entry 422 Linnaeus's *Hortus Cliffortianus*... fol. 1757
'George Clifford gave this book to Mark Catesby, and he at his death will'd it to Thos. Knowlton.'

545 Petiver's works on natural history... 2 vol. fol. 1764.
Presented to Tho. Knowlton by Emanuel da Costa, June 1765.

633 Sloane's *Natural history of Jamaica,* 2 vol.... presented to Mr Knowlton by Sir Hans Sloane. fol. 1707–25.

(1) *Gent. Mag.,* **9** (New Series), p. 544 (May 1838).

Appendix C

Some eighteenth century publications in which Thomas Knowlton's name appears as a subscriber

1. *A natural history of English insects* (1720) by Eleazar Albin.
   Included in list of subscribers, 'Mr. Thomas Knowlton, gardener'.
   At base of plate 88, 'To Tho: Knowlton gardiner at Sr Henry Penrisces, at Offly Place in Hartfordshire. This plate is humbly dedicated by Eleaz^{r}. Albin.'
2. *The natural history of birds* by Eleazar Albin.
   Included in list of subscribers in vol. **2** (1734) and vol. **3** (1738), 'Thomas Knowlton, gardiner'.
3. *The natural history of Cornwall* (1758) by William Borlase.
   Included in list of subscribers, 'Mr. Thomas Knowlton'.
4. *A philosophical account of the works of nature* (1721) by Richard Bradley.
   Included in list of subscribers, 'Thomas Knowlton, of Offley-Place, Esq.'
5. *Monasticon eboracense: and the ecclesiastical history of Yorkshire* (1758), by John Burton.
   Included in list of subscribers, 'Mr Thomas Knowlton, of Lonesburgh.'
6. *The genuine remains in verse and prose of Mr. Samuel Butler* (1759).
   Included in list of subscribers, 'Mr. Knowlton, of Londsborough, Yorkshire'.
7. *The natural history of Carolina* by Mark Catesby.
   Included in list of subscribers in vol. **1** [1730–] 1731– [1732], 'Mr. Thomas Knowlton'.
8. *Historia naturalis Testaceorum Britanniae, or, the British conchology* (1778) by Emanuel Mendes da Costa.
   Included in list of subscribers, 'Mr. Thomas Knowlton, Loansborough Castle, Yorkshire'.
9. *Drake's Eboracum: or history and antiquities of the City of York* (1736).
   Included in list of subscribers, 'Mr Knowlton of Londesburgh'.

10. *Gleanings of natural history, exhibiting figures of quadrupeds, birds, fishes, insects* by George Edwards.
    Included in list of subscribers in Part II (1760), 'Mr Thomas Knowlton of Landsbury, Yorkshire.'
11. *History of the Royal and beautiful town of Kingston-upon-Hull* (1735).
    Included in the list of subscribers, 'Thomas Knowlton'.
12. *The history of London* (1756) by William Maitland.
    Included in list of subscribers to the London edition, 'Thomas Knowlton'.
13. *The gardeners dictionary* (1731) by Philip Miller.
    Included in list of subscribers, 'Mr. Thomas Knowlton, gardener to the Right Honourable the Earl of Burlington.'
14. *The practical husbandman and planter* [n.d.] [Edited by Stephen Switzer].
    Included in list of subscribers, 'Thomas Knowlton gardener to the Right Honourable the Earl of Burlington, Lonesborough, Yorks.'
15. *Miscellaneous poems and translations* (1740) by H. Travers.
    Included in list of subscribers, 'Mr Thomas Knowlton, of Londsborough.'

Appendix D

Knowlton the man of property

Apart from his wages as gardener to the Earl of Burlington, Knowlton earned money in various other ways such as giving advice and practical help in the design and planting of gardens, and by the sale of plants. As will be noted later in referring to his will, Knowlton died a man of substance, the possessor of both property and money. The first of his transactions in the property market took place in 1737 and the last a few years before his death. Using the township index to memorials of deeds in the *East Riding Registry of Deeds,* Knowlton's interests as a land owner are found to have covered a fairly wide area. Below, references are given in a condensed form to his property in Beverley, Sculcoates, Bilton, Marfleet, Pocklington(1), Sancton, Holme upon Spalding Moor, and Shipton(2).

BEVERLEY

1. Memorial of release, 18 June 1737(3), to Thomas Knowlton of Londesborough, gent.: Walkergate garden, 2 acres.
2. Memorial of mortgage, 1 May 1755,(4) to Thomas Knowlton:
   a) house wherein Peter Hunsley dwells in Butcher Row, with yard, garden, stable, workhouse, and piece of waste ground situate in Well lane.
   b) house in Butcher Row with yard.
   c) stable, garth and gardens in Mr. Wilbert's lane.
3. Memorial of assignment, 5 May 1759,(5) to Thomas Knowlton: house wherein Peter Hunsley lately dwelt in Butcher Row, and garth, garden or yard, outhouses most used as workshops, tenement and appurtenances.

(1) Beverley, Sculcoates (now part of Hull), Bilton, Marfleet, and Pocklington are 11, 18, 19, 20, and 5 miles respectively from Londesborough.

(2) Sancton, Holme upon Spalding Moor, and Shipton, are $4\frac{1}{2}$, $51\frac{1}{2}$, and 2 miles respectively from Londesborough.

(3) Book P, p. 25 no. 66, registered 21 June 1737.

(4) Book X, p. 283 no. 649, registered 2 June 1755.

(5) Book AA, p. 442 no. 677, registered 5 May 1759.

4. Memorial of tripartite indenture, 13 October 1759,[1] transfer by Thomas Knowlton and other persons, to Suckling Spendlove of property described in 2(a)–(c).

SCULCOATES, BILTON AND MARFLEET

1. Memorial of lease and release 24/5 April 1752[2] to Thomas Knowlton:
   a) house and two closes of meadow or pasture ground, now or late used as a garden, one of the said closes commonly called or known by the name God's close – 4 acres, the other by the name of the half Moon close – 5½ acres, also all the ground upon which the Bulwarks were formerly made at the South end of the said two closes, situate within the Lordship of Sculcoates.
   b) three adjoining closes of meadow or pasture, 17 acres in all. Two closes being in Bilton and one in Marfleet.
2. Memorial of lease and release 20/21 January 1775[3] by Thomas Knowlton and wife Elizabeth to Kingston upon Hull Dock Company: property as in 1(a) ['closes . . . used as a garden'].
3. Enrolment of bargain and sale 21 January 1775[4] by Thomas Knowlton and wife Elizabeth to Dock Company, for £1,100: property as in 2.

The lease and release of 1775 is of the same Sculcoates property as in 1752, except that the closes are stated as being used as a garden.

In their work entitled *Georgian Hull* (1978/9)[5] Ivan and Elisabeth Hall publish a map of 'The Dock Company estate 1778' by Charles Tate (redrawn) which shows a plot of land, 7 acres 2 roods, referred to as 'Bourn Garden late Mr Knowlton'. 'The New Dock' cut into this plot using 2 roods for the dock itself, while the rest of the area was developed with the new road, Charlotte Street, going east-west across the centre. William Bone was the occupier in 1752; perhaps this gave rise to the name Bourn Garden.

POCKLINGTON

1. Memorial of lease and release 5/6 April 1770[6], to Thomas Knowlton: house and yard and building used as a bark mill, in or near St. Helengate.
2. Memorial of lease and release 5/6 April 1770[7], to Thomas Knowlton: house

(1) Book AB, p. 81 no. 187, registered 30 November 1759.
(2) Book W, p. 91 no. 200, registered 1 June 1752.
(3) Book AU, p. 232 no. 375, registered 25 January 1775.
(4) Book AT, p. 162 no. 18, registered 1 May 1775.
(5) p. 69.
(6) Book AM, p. 522 no. 890, registered 8 June 1770.
(7) Book AM, p. 522 no. 891, registered 8 June 1770.

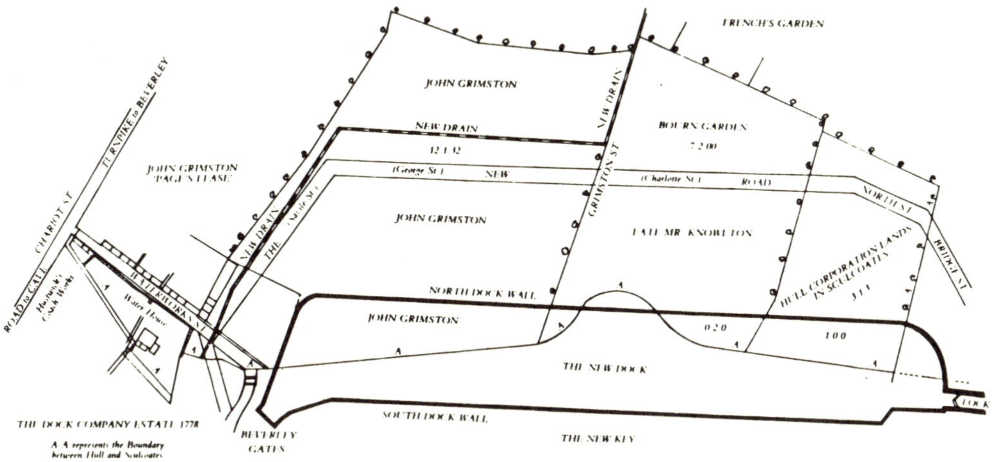

Map of Dock Company Estate 1778 by Charles Tate (redrawn), reproduced from Hall (I. and E.), Georgian Hull *(1978/9), showing the plot of land owned by Knowlton.*

or frontstead or site of same, and yard now or formerly used as a tanyard, in or near St. Helengate.

3. Memorial of release 4 October 1775(1), to Thomas Knowlton: two closes of meadow or pasture in part of Pocklington Common, called South Moor, 4 acres.
4. Memorial of lease and release 4/5 April 1777(2), to Thomas Knowlton: piece of inclosed ground near Hungate, 2 acres.

Knowlton is also mentioned:

1. In the Pocklington Inclosure Award, 18 April 1759(3)
   allotment to Thomas Knowlton of Londesborough, gardener and Mary Stephenson of Pocklington, in severality, 3 acres, 1 rood, 6 perches in [South Moor] part of Common.
2. In the Sancton Inclosure Award, 11 April 1771(4)
   3 allotments, to Thomas Knowlton as lessee of Thomas Broadley:
   25 acres, 2 roods, 32 perches.
   29 acres, 2 roods, 32 perches.
   17 acres, 3 roods, 8 perches.
   There is no description of the land itself, indicating specific land use.

(1) Book AW, p. 187 no. 310, registered 13 March 1776.
(2) Book AY, p. 216 no. 332, registered 12 June 1777.
(3) Book Y, p. 230 no. 39, registered 19 April 1759.
(4) Book AN, p. 62 no. 2, registered 12 April 1771.

3. In the Holme upon Spalding Moor Award, 1 March 1777(1)
   2 allotments to Thomas Knowlton of Londesborough, gardener:
   7 acres, 1 rood, 15 perches.
   33 acres, 3 roods, 10 perches.

There are two other references to the Holme upon Spalding Moor properties in the East Riding Deeds Register:

1. Memorial of lease and release 15/16 October 1760(2) to Thomas Knowlton, gent.:
   a) Messuage, homestead and premises with the appurtenances.
   b) garth called Hemp Garth, 2 acres.
   c) close called Fownay Close and two adjoining closes, 8 acres.
   d) $\frac{1}{2}$ oxgang of land, 9 acres(3)
2. Memorial of lease and release 1/2 June 1778(4), by Thomas Knowlton and wife Elizabeth to the Revd. William Ponsonby:
   1 acre, 1 rood, 20 perches being part of an allotment awarded upon enclosure and situate in Church Field.

SHIPTON

1. Memorial of lease and release 17/18 February 1762(5) to Thomas Knowlton: 2 oxgangs in Shipton (probably a second mortgage, discharged by certificate 5 July 1764).
2. Memorial of an indenture 18 March 1762(6) to Thomas Knowlton: cottage and garth with appurtenances in Shipton.
3. Memorial of tripartite indenture 5 July 1764(7), Knowlton and others to Richard Ryley of Hardwick, Co. of Derby: cottage with barn, stable, garth and orchard frontstead or site of Blue Stone Cottage with garth 2 oxgangs of land.

(1) Book AX, p. 357 no. 10, registered 21 April 1777.
(2) Book AB, p. 353 no. 632, registered 13 December 1760.
(3) According to Richardson (John), *The local historian's encyclopedia* (1974), Oxgang, Oxgade or Oxland is the name for a variable measure depending on soil quality. It was originally the amount of land that could be cultivated using one ox, and could vary between 7 and 50 acres. Normally it was reckoned to be one-eighth of a Hide, alternatively called a Bovate. A Hide was originally the amount of land which could be ploughed in a year using one plough and support a family. It could therefore vary in area with the soil quality – generally between 60 and 180 acres.
(4) Book AY, p. 543 no. 922, registered 18 June 1778.
(5) Book AD, p. 283 no. 585, registered 27 February 1762.
(6) Book AD, p. 297 no. 611, registered 3 April 1762.
(7) Book AE, p. 500 no. 948, registered 3 August 1764.

4. Memorial of lease and release 17/18 September 1775[1], Henry and Ann Plakston and Thomas Knowlton to William Binnington: rights over commons and open fields in Shipton appertaining to a cottage there.

Knowlton was one of the original subscribers to the Beverley – Molescroft – Kendale House and Bainton Balk Turnpike Trust. He attended the first meeting at Beverley on 1 May 1766 and advanced £100.[2]

(1) Book AV, p. 600 no. 991, registered 6 November 1775.
(2) Information from the Turnpike Trust minute-book, reference TTBM 1, preserved in Humberside County Record Office.

Appendix E

Abstract of Thomas Knowlton's will[1]

'This is the last will and testament of me Thomas Knowlton of Londsbrough in the county of York gentleman this twenty fifth day of August in the year of our Lord one thousand seven hundred and eighty. I give and devise all my messuages or dwelling-houses closes or parcels of land and hereditaments situate at Bilton Marfleet Beverley and Holme in the said County of York with all the rights privileges and appurtenances thereto belonging and also the interest or annual produce arising from the principal sum of one hundred pounds secured on Moscroft Turnpike unto my son the Reverend Charles Knowlton clerk and his assigns for and during the term of his natural life and from and after his decease I give and devise all the same hereditaments and premises together with the said one hundred pounds principal money and from thenceforth the interest thereof unto my grandson Thomas Knowlton son of the said Charles Knowlton his heirs executors administrators and assigns for ever I give and devise unto my daughter Elizabeth the wife of Robert Wilson of Pocklington in the said county of York tanner her heirs and assigns for ever all that my messuage or dwellinghouse in Saint Ellen Gate in Pocklington aforesaid with the stables tan yard garden and appurtenances thereto belonging I also give and devise unto my granddaughter Elizabeth Wilson one of the daughters of the said Robert Wilson her heirs and assigns for ever the close or croft at Pocklington aforesaid which I purchased of Mr Charles Cross and now occupied by Mr Hooper Officer of Excise I also give and devise unto my granddaughter Mary Wilson the other daughter of the said Robert Wilson her heirs and assigns for ever... my common right close at Pocklington aforesaid lately occupied by William Ireland butcher and now by William Smith I give and bequeath all my mortgages bonds promisory notes monies and securities for money goods chattels personal estate and effects whatsoever unto my said son in law Robert Wilson and my said grandson Thomas Knowlton their executors administrators and assigns upon trust in the

(1) Preserved in the Borthwick Institute of Historical Research, University of York.

first place to pay thereout all my just debts and funeral expenses together with the expenses of proving and registering this my will and also to pay the legacy hereinafter bequeathed to Mary Wharton . . . and out of the annual interest arising from the residue of my said personal estate to pay unto my dear wife Elizabeth Knowlton or her assigns during her life one annuity or clear yearly sum of forty pounds by four equal quarterly payments . . . and after payment of the said annuity of forty pounds my will is that the said Robert Wilson and Thomas Knowlton shall return to themselves the surplus of the interest and annual produce arising from my said personal estate (if any there shall be) . . . I give and bequeath unto my grandchildren the said Thomas Knowlton and to Charles Knowlton Elizabeth Knowlton and Sarah Knowlton sons and daughters of my said son Charles Knowlton the sum of one hundred pounds apiece I also give and bequeath unto my four grandchildren Elizabeth Wilson Mary Wilson Thomas Knowlton Wilson and Robert Wilson the sum of one hundred pounds apiece . . . I give and bequeath unto Mary Wharton daughter of Matthew Wharton the sum of twenty pounds . . . Lastly I appoint my said son in law Robert Wilson and grandson Thomas Knowlton executors of this my last will and testament . . .'

This will is signed and sealed by the testator, 'T. Knowlton', in the presence of three witnesses 'Hannah Adamson David Adamson John Robinson'. It was proved December 1781.

Appendix F

List of plants dealt with by Knowlton, and others, and mentioned in this book, arranged in families

ACANTHACEAE
Justicia adhatoda L.
J. hyssopifolia L.
Barleria prionitis L.

ACERACEAE
Acer campestre L.
A. pseudoplatanus L.

AGAVACEAE
Agave americana L. cv. Variegata
Dracaena draco L.
Sansevieria guineensis (L.) Willd.
S. zeylanica (L.) Willd.
Yucca filamentosa L.

AÏZOACEAE
Drosanthemum micans (L.) Schwantes
D. hispidum (L.) Schwantes
Mesembryanthemum crystallinum L.
Oscularia deltoides (L.) Schwantes

ALLIACEAE
Agapanthus africanus (L.) Hoffsgg.

ALSTROEMERIACEAE
Alstroemeria pelegrina L.

ALTINGIACEAE
Liquidambar styraciflua L.

AMARANTHACEAE
Amaranthus cruentus L.
Gomphrena globosa L.

AMARYLLIDACEAE
Crinum zeylanicum (L.) L.
Galanthus nivalis L.
Hippeastrum reginae Herbert
Narcissus pseudonarcissus L.
Nerine sarniensis (L.) Herbert

ANACARDIACEAE
Pistacia lentiscus L.

ANNONACEAE
Annona reticulata L.
Asimina triloba (L.) Dunal

APOCYNACEAE
Catharanthus roseus (L.) G. Don
Ichnocarpus frutescens (L.) R. Br.
Nerium indicum Miller
N. oleander L.

AQUIFOLIACEAE
Ilex cassine L.
I. vomitoria Aiton

ARACEAE
Colocasia esculenta (L.) Schott
Typhonium trilobatum (L.) Schott

ARALIACEAE
Aralia spinosa L.
Hedera helix L.

ARISTOLOCHIACEAE
Asarum europaeum L.

ASPLENIACEAE
Asplenium viride Hudson

ATHYRIACEAE
Woodsia alpina (Bolton) Gray

BERBERIDACEAE
Epimedium alpinum L.

BETULACEAE
Alnus glutinosa (L.) Gaertner cv. Rubrinervia
Betula nana L.

BIGNONIACEAE
Crescentia cujete L.

BIXACEAE
Bixa orellana L.

BORAGINACEAE
Myosotis sylvatica Hoffm.

BROMELIACEAE
Ananas cv.
Ananas comosus (L.) Merr.

CACTACEAE
Acanthocereus tetragonus (L.) Hummelinck
Mammillaria simplex Haw.
Melocactus communis Link & Otto
Opuntia ficus-indica (L.) Miller
O. tuna (L.) Miller.

CALYCANTHACEAE
Calycanthus floridus L.

CAMPANULACEAE
Campanula latifolia L.
C. medium L.
C. persicifolia L.
C. trachelium L.
Canarina canariensis (L.) Vatke
Legousia pentagonia (L.) Druce
Trachelium caeruleum L.

CANELLACEAE
Canella winteriana (L.) Gaertner

CAPPARIDACEAE
Crateva tapia L.

CAPRIFOLIACEAE
Lonicera periclymenum L.
L. sempervirens L.
Viburnum lantana L.
V. opulus L.
V. tinus L.

CARICACEAE
Carica papaya L.

CARYOPHYLLACEAE
Honkenya peploides (L.) Ehrh.

CELASTRACEAE
Maytenus lucida (L.) Loes.

CHENOPODIACEAE
Atriplex hortensis L.
Beta vulgaris L. ssp. *maritima* (L.) Arcangeli
Salicornia europaea L.
Salsola kali L.

CISTACEAE
Helianthemum nummularium (L.) Miller

CNEORACEAE
Cneorum tricoccum L.

COMMELINACEAE
Commelina africana L.

COMPOSITAE
Anthemis cotula L.
Arctotis aspera L.
Artemisia arborescens L.
Aster amellus L.
Baccharis halimifolia L.
Callistephus chinensis (L.) Nees
Carthamus arborescens L.
Centaurea cineraria L.
Chrysocoma coma-aurea L.
C. tridentata DC.
Helichrysum orientale (L.) DC.
Inula candida (L.) Cass.
Kleinia neriifolia Haw.
Osteospermum fruticosum (L.) Norlindh
Senecio doria L.
S. papillaris (L.) Jacobs
Tagetes minuta L.

CONVOLVULACEAE
Convolvulus canariensis L.
C. cneorum L.

CORNACEAE
Cornus suecica L.

CORYLACEAE
Carpinus betulus L.
Corylus avellana L.

CRASSULACEAE
Sedum dasyphyllum L.
Umbilicus rupestris (Salisb.) Dandy

CRUCIFERAE
Cakile maritima Scop.
Cardamine impatiens L.
C. trifolia L.
Iberis semperflorens L.

CRYPTOGRAMMATACEAE
Cryptogramma crispa (L.) R.Br. ex Hook.

CUCURBITACEAE
Citrullus lanatus (Thunb.) Matsum. & Nakai

CUPRESSACEAE
Cupressus lusitanica Miller
Juniperus sabina L.

DICKSONIACEAE
Cibotium barometz J. Smith

DROSERACEAE
Dionaea muscipula Ellis

EBENACEAE
Euclea racemosa Murray

ELAEAGNACEAE
Hippophaë rhamnoides L.

EMPETRACEAE
Empetrum nigrum L.

EPHEDRACEAE
Ephedra distachya L.

ERICACEAE
Andromeda polifolia L.
Arbutus andrachne L.
Arctostaphylos uva-ursi (L.) Sprengel
Rhododendron maximum L.
Vaccinium oxycoccos L.
V. vitis-idaea L.

EUPHORBIACEAE
Clutia alaternoides L.
C. pulchella L.
Euphorbia antiquorum L.
E. cyparissias L.
Pedilanthus tithymaloides (L.) Poiteau

FAGACEAE
Fagus sylvatica L.
Quercus ilex L.
Q. suber L.

FLACOURTIACEAE
Samyda dodecandra Jacq.

FRANKENIACEAE
Frankenia laevis L.

FUMARIACEAE
Dicentra cucullaria (L.) Bernh.

GERANIACEAE
Erodium maritimum (L.) L'Hér.
Geranium versicolor L.

GENTIANACEAE
Gentiana pneumonanthe L.
Gentianella amarella (L.) Börner

GRAMINEAE
Arundo donax L. var. *versicolor* (Miller) Stokes
Glyceria fluitans (L.) R.Br.
G. maxima (Hartman) Holmberg
Oryza sativa L.
Saccharum officinarum L.

GUTTIFERAE
Garcinia mangostana L.

HAMAMELIDACEAE
Hamamelis virginiana L.

HERNANDIACEAE
Hernandia sonora L.

HYDROCHARITACEAE
Stratiotes aloides L.

IRIDACEAE
Romulea columnae Sebast. & Mauri
Tigridia pavonia (L.f.) DC.

LABIATAE
Ballota africana (L.) Benth.
Cedronella canariensis (L.) Webb & Berth.
Glechoma hederacea L.
Origanum dictamnus L.
O. majorana L.
O. sipyleum L.
Pycnanthemum virginianum (L.) Pers.
Teucrium marum L.

LAURACEAE
Cinnamomum camphora (L.) J. S. Presl
C. verum J. S. Presl.
Persea americana Miller
P. indica (L.) Sprengel
Sassafras albidum (Nutt.) Nees

LEGUMINOSAE
Abrus precatorius L.
Acacia cornigera (L) Willd.
A. nilotica (L.) Willd. ex Del.
Astragalus danicus Retz.
Cajanus cajan (L.) Huth
Calliandra houstonii (Miller) Standley
Cassia bicapsularis L.
Cercis siliquastrum L.

Dorycnium pentaphyllum Scop.
Genista hispanica L.
G. libanotica Boiss.
Haematoxylum campechianum L.
Lathyrus latifolius L.
L. odoratus L.
Lotus jacobaeus L.
Medicago intertexta (L.) Miller
Mimosa sensitiva L.
Robinia pseudoacacia L.
Spartium junceum L.
Tamarindus indica L.

LENTIBULARIACEAE
Utricularia vulgaris L.

LILIACEAE
Aloe humilis (L.) Miller
A. variegata L.
Asparagus officinalis L.
Asphodelus fistulosus L.
Eucomis regia (L.) L'Hér.
Gloriosa superba L.
Haworthia retusa (L.) Haw.
Narthecium ossifragum (L.) Hudson
Tulipa cv. Duke van Tholl
T. cv. Cleremont

MAGNOLIACEAE
Liriodendron tulipifera L.
Magnolia acuminata L.
M. grandiflora L.
M. tripetala L.
M. virginiana L.

MALVACEAE
Abutilon indicum (L.) Sweet
Althaea officinalis L.
Gossypium arboreum L.
Hibiscus mutabilis L.
Malva verticillata L. var. *crispa* L.

MARANTACEAE
Maranta arundinacea L.

MARTYNIACEAE
Martynia annua L.

MELIACEAE
Melia azedarach L.
Swietenia mahagoni (L.) Jacq.

MELIANTHACEAE
Melianthus major L.
M. minor L.

MENISPERMACEAE
Menispermum canadense L.

MENYANTHACEAE
Nymphoides peltata (S. Gmelin) O. Kuntze

MORACEAE
Artocarpus heterophyllus Lam.
Broussonetia papyrifera (L.) Vent.
Ficus carica L.
F. nymphaeifolia Miller
F. religiosa L.

MUSACEAE
Musa cv.
M. x *paradisiaca* L.

MYRICACEAE
Myrica gale L.

MYRTACEAE
Myrtus communis L.
Pimenta dioica (L.) Merr.
Psidium guajava L.

NYMPHAEACEAE
Nuphar lutea (L.) Smith
Nymphaea alba L.
N. tuberosa Paine

OLEACEAE
Fraxinus excelsior L
Jasminum fluminense Vellozo
J. grandiflorum L.
J. humile L.
J. sambac (L.) Aiton

ONAGRACEAE
Epilobium angustifolium L.

OPHIOGLOSSACEAE
Botrychium lunaria (L.) Sw.

ORCHIDACEAE
Cephalanthera damasonium (Miller) Druce
Coeloglossum viride (L.) Hartman
Cypripedium calceolus L.
Himantoglossum hircinum (L.) Sprengel
Neottia nidus-avis (L.) Rich.
Ophrys apifera Hudson

OROBANCHACEAE
Lathraea squamaria L.
Orobanche hederae Duby

PASSIFLORACEAE
Passiflora laurifolia L.

PEDALIACEAE
Sesamum orientale L.

PHILADELPHACEAE
Philadelphus coronarius L.

PHYTOLACCACEAE
Rivina humilis L.

PINACEAE
Larix decidua Miller
Picea abies (L.) Karsten
Pinus sylvestris L.

PIPERACEAE
Piper amalago L.

PLANTAGINACEAE
Plantago major L.
P. major L. forma *ramosa* Pilger

PLUMBAGINACEAE
Armeria maritima Miller
Limonium vulgare (Miller) Willd.

POLYGALACEAE
Muraltia heisteria (L.) DC.

POLYGONACEAE
Oxyria digyna (L.) Hill
Polygonum undulatum (L.) P. Bergius
Rheum ribes L.

PORTULACACEAE
Claytonia virginica L.

PRIMULACEAE
Anagallis monelli L.
Cortusa matthioli L.
Cyclamen persicum Miller
C. vernale Miller
Glaux maritima L.
Primula auricula L.
P. farinosa L.
Soldanella alpina L.
Trientalis europaea L.

PUNICACEAE
Punica granatum L.

PYROLACEAE
Pyrola rotundifolia L.

RANUNCULACEAE
Actaea spicata L.
Adonis vernalis L.
Anemone apennina L.

Clematis virginiana L.
Eranthis hyemalis (L.) Salisb.
Knowltonia capensis (L.) Huth
Ranunculus ficaria L.

ROSACEAE
Crataegus monogyna Jacq.
C. monogyna Jacq. cv. Biflora
Fragaria chiloensis (L.) Duchesne
Mespilus germanica L.
Prunus lusitanica L.
P. padus L.
Rosa cinnamomea L.
R. *foetida* J. Herm.
R. *gallica* L. 'Violacea'
R. *moschata* J. Herm.
Rubus chamaemorus L.
R. *saxatilis* L.
Sorbus aria (L.) Crantz
S. aucuparia L.

RUBIACEAE
Anthospermum aethiopicum L.
Cephaëlis ipecacuanha (Brot.) Tussac
Coffea arabica L.
Gardenia jasminoides Ellis
Phyllis nobla L.

RUTACEAE
Citrus aurantium L.
C. grandis (L.) Osbeck
C. limon (L.) Burm. f.
C. medica L.
C. paradisi Macfad.
C. sinensis (L.) Osbeck

SALICACEAE
Populus alba L.
P. balsamifera L.
Salix viminalis L.

SAPOTACEAE
Chrysophyllum cainito L.

SARRACENIACEAE
Sarracenia purpurea L.

SCROPHULARIACEAE
Asarina procumbens Miller
Digitalis purpurea L.
Halleria lucida L.
Isoplexis canariensis (L.) Loudon
Linaria aeruginea (Gouan) Cav.

SOLANACEAE
Capsicum frutescens L.
Datura metel L.
Physalis alkekengi L.

STERCULIACEAE
Hermannia grossularifolia L.

STYRACACEAE
Halesia carolina L.

TAXACEAE
Taxus baccata L.

TETRAGONIACEAE
Tetragonia fruticosa L.

THEACEAE
Camellia sinensis (L.) O. Kuntze

TILIACEAE
Tilia x *europaea* L.

TRILLIACEAE
Paris quadrifolia L.

TROPAEOLACEAE
Tropaeolum majus L.

ULMACEAE
Celtis australis L.
Ulmus procera Salisb.

UMBELLIFERAE
Crithmum maritimum L.
Eryngium maritimum L.
Ferula communis L.
Myrrhis odorata (L.) Scop.

VERBENACEAE
Verbena hastata L.
Vitex agnus-castus L.

VISCACEAE
Viscum album L.

List of algae and fungi dealt with by Knowlton

ALGAE
Cladophora aegagropila (L.) Rabenh.
Himanthalia elongata (L.) Gray
Palmaria palmata (L.) O. Kuntze

FUNGI
Lactarius vellerius (Fries) Fries
L. piperatus (Scop. ex Fries) Gray

Sources of illustrations

Indexes

General index

Note: Only the more significant personal names are entered; personal acknowledgements and references to the literature are excluded. Page numbers in italic type refer to illustrations; inclusive page numbers followed by *passim* indicate scattered references; page numbers followed by *n* refer to footnotes. Plant species are indexed separately on pp. 318–324

Plant index

Note: Plant species have been entered separately under both Latin and common names where both names are supplied by the text. Page numbers in italic type refer to illustrations; inclusive page numbers followed by *passim* indicate scattered references; page numbers followed by *n* refer to footnotes. Pre-Linnaean Latin plant names have not been entered.